CURRENTS OF THE UNIVERSAL BEING

CURRENTS OF THE UNIVERSAL BEING

explorations in the literature of energy

Scott Slovic, James E. Bishop, and Kyhl Lyndgaard

Texas Tech University Press

This book is typeset in Minion Pro. The paper used in this book meets the minimum requirements of ANSI/NISO Z39.48-1992 (R1997). ♾

Designed by Ryan Miller
Cover designed by Ryan Miller

Library of Congress Cataloging-in-Publication Data

Currents of the universal being : explorations in the literature of energy / edited by Scott Slovic, James E. Bishop, and Kyhl Lyndgaard.
pages cm
Summary: "Presents a variety of poetry, fiction, nonfiction, and interviews concerning energy and energy related issues. Aspires to place the technical subject of energy into a broader cultural context"—Provided by publisher.
Includes bibliographical references and index.
ISBN 978-0-89672-928-5 (paperback)—ISBN 978-0-89672-929-2 (e-book) 1. Power resources. 2. Science—Literary collections. 3. American literature—21st century. 4. Nature in literature. I. Slovic, Scott, 1960– II. Bishop, James E. III. Lyndgaard, Kyhl.
PS509.S3C87 2015
810.8'0356—dc23

2015001545

Printed in United States
15 16 17 18 19 20 21 22 23 / 9 8 7 6 5 4 3 2 1

Texas Tech University Press
Box 41037
Lubbock, Texas 79409-1037 USA
800.832.4042
ttup@ttu.edu
www.ttupress.org

Contents

Acknowledgments

This book emerges from two beliefs: that energy is one of the world's crucial environmental and social issues and that literature (and other cultural media) are vital ways of exploring the meaning of the phenomenon of energy. We offer this collection as a means of supplementing readers' thinking about energy through more conventional disciplinary routes, such as economics, law, natural resource sciences, and engineering. There are also aesthetic, philosophical, and social justice dimensions that need to be recognized when we talk about energy.

Our own thinking about the cultural—and especially the *literary*—aspects of energy took focus during the fall of 2006 when Scott Slovic offered a graduate seminar on "The Literature of Energy" at the University of Nevada, Reno, where he was teaching at the time. Jim Bishop and Kyhl Lyndgaard were then doctoral students and participated in this seminar. We would like to acknowledge the other members of the seminar, all of whom contributed to this book as it germinated: Megan Kuster, Anna McCarthy, Deidre Pike, Eric Stottlemyer, and Diana Villanueva Romero.

As the project evolved, Scott moved to the University of Idaho, Jim took teaching positions first at Young Harris College in Georgia and later at the Colorado School of Mines, and Kyhl taught at Luther College in Iowa before moving to his current position at Vermont's Marlboro College. We are grateful for the support we have received from all of our institutions. In particular, a research grant from the University of Nevada, Reno's Academy for the Environment provided essential funding, without which this book could not have taken shape. Luther College, with the Associated Colleges of the Midwest and the Mellon Foundation, also offered support in the form of a research grant, as did Marlboro College. Luther College student worker Katelyn Van Wyck assisted in the compilation of the manuscript.

We derive energy in our lives not only from the natural world and through the technologies designed to harvest and channel such forces, but from the people and environments that inspire us. Scott would like to thank Susie Bender, as always, for her avid interest in his many ecocritical projects—and Hanna for running alongside him during the final years of this editorial work. Jim is grateful to Rhiannon McClatchey for her enthusiasm for this project and her generosity in contributing a piece of her own original work. He would also like to thank his parents, Jim and Peggy Bishop, for their unwavering support through the long gestation of this manuscript. Kyhl thanks Marian Lyndgaard for her encouragement and support throughout the project, especially since the arrival of Lars and David, who possess and share original energies all their own.

We gratefully acknowledge the following authors, agents, and publishers for permission to reprint the work included in this anthology:

A. R. Ammons, "Mechanism," from *Collected Poems: 1951–1971*. Copyright © 1960 by A. R. Ammons. Used by permission of W. W. Norton & Company, Inc.

Marilou Awiakta, "Baring the Atom's Mother Heart," from *Selu: Seeking the Corn-Mother's Wisdom*. Copyright © 1993 by Marilou Awiakta. Reprinted with the permission of Fulcrum Publishing.

Rick Bass, excerpt from "A Short History of Montana," from *The Heart of the Monster* by Rick Bass and David James Duncan (Missoula: All Against the Haul, 2010). Permission granted by the author.

Dan Bloom, "To heal the world, to 'repair the world.'" Reprinted by permission of author.

John Calderazzo, "Into the Ring of Fire," from *Rising Fire: Volcanoes and Our Inner Lives*. Copyright © 2004 by John Calderazzo. Reprinted with the permission of the author and The Lyons Press/Rowman & Littlefield Publishing Group.

Alfred W. Crosby, "The Largess of the Sun," from *Children of the Sun: A History of Humanity's Unappeasable Appetite for Energy*. Copyright ©

2006 by Alfred W. Crosby. Used by permission of W. W. Norton & Company, Inc.

Joan Didion, "At the Dam," from *The White Album*. Copyright © 1979 by Joan Didion. Reprinted by permission of the author and Farrar, Straus & Giroux, LLC.

Ricardo García, "Sign at the Mine: No Women or Bears Allowed," from *Coal Camp Days: A Boy's Remembrance*. Copyright © 2001 by University of New Mexico Press. Reprinted with permission.

David Gessner, "Energy," from *The Tarball Chronicles: A Journey beyond the Oiled Pelican and into the Heart of the Gulf Oil Spill*. Copyright © 2001 by David Gessner. Reprinted with permission from Milkweed Editions, www.milkweed.org, and by Scovil Galen Ghosh Literary Agency, Inc.

Drum Hadley, "A Jack of All Trades," from *Voice of the Borderlands*. Copyright © 2005 by Drum Hadley. Reprinted with the permission of Rio Nuevo Publishers.

William Heyen, "Pterodactyl Rose," from *Pterodactyl Rose: Poems of Ecology* (St. Louis, Missouri: Time Being Books, 1991). Copyright © 1991 by Timeless Press, Inc. Reprinted with the permission of the author.

Marybeth Holleman, "What Happens When Polar Bears Leave?" from *ISLE: Interdisciplinary Studies in Literature and Environment* (Summer 2007). Reprinted with the permission of the author.

Robinson Jeffers, "O Lovely Rock," from *The Collected Poetry of Robinson Jeffers: Volume 2, 1928–1938*. Copyright © 1938 by Robinson Jeffers, renewed 1966 and copyright © Garth and Donnan Jeffries. Reprinted with the permission of Stanford University Press.

David Lee, "Remnant," from *Stone Wind Water: Poems*. Copyright © 2010 by David Lee. Reprinted with the permission of the author and Black Rock Press.

Audre Lorde, "Coal," from *The Collected Poems of Audre Lord*. Copyright © 1997 by The Audre Lorde Estate. Used by permission of W. W. Norton & Company, Inc., and Charlotte Sheedy, literary agent.

Adrian C. Louis, "Nevada Red Blues," from *Among the Dog Eaters*. Copyright © 1992 by Adrian C. Louis. Reprinted with the permission of The Permissions Company, Inc., on behalf of West End Press, www.westendpress.org.

Jerry Martien, "Now the Ice." First published as a broadside by Tangram Press, Berkeley, 2006. Reprinted with the permission of the author.

Cate Marvin, "A Windmill Makes a Statement," from *Fragment of the Head of a Queen*. Copyright © 2007 by Cate Marvin. Reprinted with the permission of the author and with the permission of The Permissions Company, Inc., on behalf of Sarabande Books, www.sarabandebooks.org.

David Mas Masumoto, "Callous Hands," from *Harvest Son: Planting Roots in American Soil*. Copyright © 1998 by David Mas Masumoto. Used by permission of W. W. Norton & Company, Inc.

Rhiannon McClatchey, "Kinesis." Used with the permission of the author.

Don McKay, "Nocturne Macdonald Cartier Freeway," from *Camber: Selected Poems*. Copyright © 2004 by Don McKay. Reprinted with the permission of Random House of Canada Limited.

Bill McKibben, "A Moral Atmosphere," from *Orion* (March/April 2013). Copyright © 2013. Reprinted by permission of author.

Susan McLean, "Reaping the Wind," from *First Things* (August/September 2010). Copyright © 2010. Reprinted with the permission of Susan McLean.

Gary Paul Nabhan, excerpt from *Coming Home to Eat: The Pleasure and Politics of Local Food*. Copyright © 2002 by Gary Paul Nabhan. Used by permission of W. W. Norton & Company, Inc.

Marilyn Nelson, "Minor Miracle," from *The Fields of Praise: New and Selected Poems*. Copyright © 1997 by Marilyn Nelson. All rights reserved. Reprinted with permission of Louisiana State University Press.

Simon Ortiz, "Electric Lines" and "Gas Lines," from *Woven Stone* (Tucson: University of Arizona Press, 1992). Copyright © 1992 by Simon Ortiz. Reprinted with the permission of the author.

E. C. Pielou, "Energy Is Everywhere," from *The Energy of Nature*. Copyright © 2001 by E. C. Pielou. Reprinted with the permission of the author and by the University of Chicago Press.

Janisse Ray, "First Landfall," from *UnspOILed: Writers Speak for Florida's Coast*, edited by Susan Cerulean, Janisse Ray, and A. James Wohlpart (Tallahassee, Florida: Heart of the Earth/Gandy Printers, 2010). Reprinted with the permission of the author.

Pattiann Rogers, "The Power of the Sun," from *Firekeeper: New and Selected Poems*. Copyright © 1994 by Pattiann Rogers. Reprinted with permission from Milkweed Editions, www.milkweed.org.

Wendy Rose, "Plutonium Vespers," from *Bone Dance: New and Selected Poems, 1965–1993*. Copyright © 1994 by The Arizona Board of Regents. Reprinted with the permission of the University of Arizona Press.

Muriel Rukeyser, "Time Hinder Not Me, His Arms Reach Here and There," from *The Collected Poems of Muriel Rukeyser*. Copyright © 1957 by Muriel Rukeyser. Reprinted with the permission of International Creative Management, Inc. All rights reserved.

John C. Ryan and Alan Thein Durning, "Bike (and Car)," from *Stuff: The Secret Lives of Everyday Things*. Copyright © 1997 by Northwest Environment Watch. Reprinted with permission.

Gary Snyder, "Oil," from *The Back Country*. Copyright © 1968 by Gary Snyder. Reprinted by permission of New Directions Publishing Corp.

Rebecca Solnit, "April Fool's Day," from *Savage Dreams: A Journey into the Hidden Wars of the American West*. Copyright © 1994 by Rebecca Solnit. Reprinted with the permission of the University of California Press.

William Stafford, "Maybe Alone on My Bike," from *The New Yorker* (April 4, 1964). Copyright © 1964 by William Stafford. Reprinted with the permission of The Permissions Company, Inc., on behalf of the William Stafford Family Trust.

Sandra Steingraber, "Shale Game," from *Orion* (May/June 2010). Reprinted with the permission of the author.

Sheryl St. Germain, "Midnight Oil." Originally published in *Navigating Disaster: Sixteen Essays of Love and a Poem of Despair* (Louisiana Literature, 2012). Reprinted with the permission of the author.

John Updike, "Energy: A Villanelle," from *Collected Poems, 1953–1993*. Copyright © 1993 by John Updike. Used by permission of Penguin Group (UK) Ltd. and Alfred A. Knopf, a division of Random House, LLC.

Introduction

. . . the currents of the Universal Being circulate through me. . . .

—Ralph Waldo Emerson, *Nature* (1836)

Tug at any human use of energy and you will find its effects cascading throughout society, spilling into the environment and coming back to us. As we were building the edifice of the first high-energy society many things got unraveled in the process but one key reality made the task easier: during the twentieth century we were largely on a comfortable, and a fairly predictable, energy path of a mature fossil-fuel civilization. Things are different now: the world's energy use is at the epochal crossroads. The new century cannot be an energetic replica of the old one and reshaping the old practices and putting in place new energy foundations is bound to redefine our connection to the universe.

—Vaclav Smil, *Energy at the Crossroads: Global Perspectives and Uncertainties* (2003; 373)

The concept of energy underlies all contemporary environmental issues, but the depth and pervasiveness of energy go well beyond environmental concerns and consciousness. In a philosophical and biological sense, life itself is an expression of energy. Our perspective, in compiling this anthology, is flexible and encompassing. Rather than simply providing readings that steer audiences toward the fleeting questions of the day—Is the cost of a barrel of oil rising or falling? What are the most viable current alternatives to fossil fuels?—we want readers to engage with the old and new writings collected here in ways that deepen their appreciation for the profundity of "energy" as an idea. Entire philosophies of human experience are linked to the notion of energy as a pervasive life force. Several years ago, a visiting poet from South Korea made a memorable impact on the University of Nevada English Department when he began practicing "qi kung" on behalf of ailing colleagues (literally, pressing cosmic energy into aging backs through his thumbs!). At first glance, North Americans are likely to view the medical encouragement of "qi" (or "energy flow"), the underlying principle of Chinese medicine, as a quaintly exotic notion of how the human body works. We'd rather pop pills. But alternative medical traditions, including some from South and East Asia, are now sweeping the United States. And even if we squirm at the idea of minding our bodily qi, Americans are quite likely to accept Emerson's formulation in the epigraph above, sensing that something powerful is passing through us and all other living beings, something that gives us animation. We might well call this "current," whether electrical or spiritual or both, "energy."

Perhaps because energy is a fundamental aspect of our daily lives—from the homes we heat, to the cars we drive, to the foods we eat, to the very blood coursing through our veins and the ideas through our gray matter—it is also something that we often take for granted. Electricity and other forms of energy are almost always available in urban North America and Europe and in other industrialized parts of the world, and although everyone uses energy, few understand its origins, its complexities, or its limitations. Government reports, NGO assessments, newspapers, magazines, and scientific journals constitute the forums in which we expect to find contemporary discussions of energy. However, cultural meaning is manifested in our physical and psychological relationships to various forms of energy, and this meaning finds powerful expression in fiction, literary essays, and poetry. This anthology of energy literature seeks to provide reflective insight into this critically important topic across broad spectrums of style, genre, and interpretation. We all have an important stake in understanding the meaning of energy—spiritual energy, bodily energy, electric power, the use of fossil and alternative fuels, and so forth—in our lives. The literature of energy is a powerful tool to guide our thinking about the phenomenon of energy, a means of helping all readers appreciate how the "currents of the Universal Being" (energy in its many incarnations) pass through our individual and collective lives.

Divided into three sections and eight chapters, this anthology incorporates a wide range of literature about energy, from the esoteric to the technical, from the lyrical to the practical. The first section, *Living With (and Without) Energy*, includes texts that define energy and its impacts on human and nonhuman life, exploring the contemporary parameters of energy use and abuse. Section two, *Energy Debates*, investigates the physical energy that powers human existence and contrasts fossil fuels with nuclear power and alternative resources. It redirects the debate back to the earth, exploring the environmental consequences—both negative and positive—of the human use of energy in its multifarious forms. The third section offers additional reading suggestions as well as a range of syllabi. Our intention is for this book to be a useful class text and jumping-off point for more sustained study of the topic. Although we attempt to offer a flexible and encompassing range of energy-related topics in this book, we have not sought to be fully encyclopedic, in part because there are not anthologizable literary texts pertinent to every possible form of energy and also because we hope to engage some of the most significant forms of energy more deeply by providing focused coverage in this collection.

In the United States, we became acutely conscious of energy as a "contemporary issue" in the 1970s when certain global events called into question the possibility of ample energy resources in perpetuity. When suppliers of oil in the Middle East clamped down on the production of oil in 1973, following the Yom Kippur War, American consumers found themselves lining up around the block to fill their cars with expensive gasoline—and a new consciousness of energy scarcity began to emerge. Six years later, on July 15, 1979, President Jimmy Carter called for a new reduction in oil imports and attention to energy efficiency. One of the little-known accomplishments of the Carter presidency was the installation of solar panels on the White House as both a symbolic and practical gesture of concern for energy consumption in the United States. Shortly after taking office in 1981, though, President Ronald Reagan had the panels removed from the White House, suggesting an optimistic, but likely short-sighted, view of the country's energy future. Some panels from the White House were acquired in 1991 by environmentally oriented Unity College in Maine and were installed on the school's cafeteria. In 2010, a delegation from Unity led by climate change activist Bill McKibben traveled to Washington in an attempt to return the panels to President Barack Obama for re-installation on the White House. While this specific gift was declined, Obama did pledge to have modern panels installed. The new panels finally appeared on the roof of the White House in May 2014, symbolizing the American government's renewed commitment to energy efficiency and renewable energy sources.

During the heyday of American energy consciousness, Wendell Berry published his classic essay titled "The Use of Energy" (in his 1977 book *The Unsettling of America: Culture & Agriculture*), observing that "[t]he energy crisis is not a crisis of technology but of morality." "The issue," he wrote, "is restraint. The energy crisis reduces to a single question: Can we forbear to do anything that we are able to do?" (94–95). In some ways, this question also asks that we, in the twenty-first century, take a moment to unplug ourselves from smart phones and computerized pads to imagine what it means to be plugged into the world itself through our natural senses. Might there be some virtue to "social networking," with our own kind and with the larger planet, that goes beyond the Internet? Might we somehow benefit from *not* turning on our devices, at least some of the time? Lest this seem like a grim and idle vision of an unnecessarily boring reality, consider historian Albert W. Crosby's concluding chapter in *Children of the Sun: A History of Humanity's Unappeasable Appetite for Energy* (2006), in which he cautions readers to keep in mind the possibility that our energy history may not be a perpetual process of ramping up. "We children of the sun may be standing on the peak of our energy achievements poised for the next quantum leap upward," he writes. "Or we may be teetering there, destined to participate in nature's standard operational procedure of pairing a population explosion with a population crash" (164). Recalling a sweeping electrical blackout that briefly paralyzed several major U.S. and Canadian cities in August 2003, Crosby noted, "For many people, the brief loss of energy being delivered from the Sun as electricity turned out to be more of an adventure than a disaster. For others, the blackout of 2003 was a premonitory vision, acutely foreshortened to enhance comprehension, of our possible course in the next few centuries" (166). These are the final words of his book. The upshot is that it behooves us to think deeply about the role of energy in every aspect of our lives and to keep an open mind about what it would mean if we suddenly or gradually found ourselves flipping electrical switches and remaining in darkness. Writers and sustainability experimenters, such as Colin Beavan, the author of *No Impact Man* (2009), have begun exploring the post-electrical and post-oil possibilities of our society.

As the global human population swells beyond seven billion people, and as developed and developing countries consume ever-greater amounts of the world's resources, energy conservation is arguably the most critical issue of the twenty-first century. Canadian scholar Vaclav Smil suggested this when he wrote, in his 2003 volume *Energy at the Crossroads*, "Tug at any human use of energy and you will find its effects cascading throughout society" (373). Current rates of consumption presuppose a limitless supply of the world's energy resources, an attitude that sets the stage for energy conflicts whose consequences will lead either to technological innovation and resource protection or environmental collapse and territorial warfare. Solutions to these problems will not be simple. As global temperatures continue to rise, as fuel supplies dwindle, and as our dependence on foreign energy sources threatens national security (in the United States and elsewhere), people will increasingly need ways to talk about energy. Contrasting ideas about energy use need to be discussed openly and safely.

The literature of energy helps people to engage with a changing world and provides a medium through which readers can reflect critically upon complicated energy issues whose solutions will inevitably involve personal (and societal) sacrifices and transformations. With this anthology, we have aimed to assemble a suggestive compendium of texts to stimulate discussions about energy consumption, to reflect upon and explore energy but not to reduce opinions into simple polemics about lifestyle choices. Although we think of this volume principally as a textbook for college-level writing and sustainability-studies courses, we believe other readers will also enjoy and learn from the material we've gathered here.

Works Cited

Berry, Wendell. "The Use of Energy." *The Unsettling of America: Culture & Agriculture*. New York: Avon, 1977. 81–95.

Crosby, Alfred W. *Children of the Sun: A History of Humanity's Unappeasable Appetite for Energy*. New York: Norton, 2006.

Emerson, Ralph Waldo. *Nature*. 1836. *Nature/Walking*. Introduction by John Elder. Boston: Beacon, 1994.

Smil, Vaclav. *Energy at the Crossroads: Global Perspectives and Uncertainties*. Cambridge, MA: MIT P, 2003.

SECTION ONE

living with (and without) energy

Introduction to Living With (and Without) Energy

People have harnessed energy from the sun, wind, water, and fire for tens of thousands of years for diverse purposes such as transportation, cooking, and altering their environment. Only in the last two centuries, however, have we begun wide-scale usage of nonrenewable fossil fuels. Loren Eiseley, in his essay "Man the Firemaker" (1954), suggests that humans are "a flame—a great, roaring, wasteful furnace devouring vast irreplaceable substances of the earth." Eiseley looks to sustain the high global population that was made possible through the exploitation of fossil fuels for energy. With the advent of the electrical grid across North America, Europe, and now much of the world, people increasingly take cheap, plentiful, and instantaneous energy for granted. "Living With (and Without) Energy" offers many deliberate treatments of energy usage, a subject which is of great importance as those "vast irreplaceable substances" dwindle.

Many of the poets and writers in this section suggest that, once started, energy projects and the associated technology take on an irresistible life of their own. For instance, Gary Snyder's poem "Oil" considers an oil tanker, concluding that its contents are what "these / crazed, hooked nations need: / steel plates and / long injections of pure oil." Others take a more celebratory stance, expressing gratitude for the conveniences of mass transit, or even joy while harnessing and controlling vast amounts of energy made visible through electric light.

"Reflections Upon the Grid" is an examination of the essential nature of energy. The first law of thermodynamics states that energy can be changed from one form to another, but cannot be created or destroyed. Many selections in this opening chapter show social and structural ramifications of the abundant energy that is delivered to affluent populations while "downstream" populations deal with the cost of that energy in the form of emissions and infrastructure. Other selections show the ways individuals can, at times, exert choice over their interactions with modern energy. Regardless of our personal or social circumstances, however, Alfred W. Crosby writes in universal terms about the fundamental relationship between living organisms and energy: "All humans and all organisms are dependent on external sources for fuels. All are parasitic."

Early muscle- and wind-powered boats and sleds have long since been replaced by machines powered by internal combustion engines, and " 'A Going Thing': Machines, Mountains, and Muscles" explores the relationship that people have with these shifts in transportation options created with the help of concentrated sources of energy. For example, Sinclair Lewis, in *Main Street* (1920), suggests that the train was viewed as a "new god; a monster of steel limbs, oak ribs, flesh of gravel, and a stupendous hunger for freight; a deity created by man." While not all selections are as metaphorical as Lewis's Midwestern train, the variety of ways that mobility—be it by road, rail, or air—has been anthropomorphized and romanticized is stunning.

Human beings are not only exploiters of energy derived from nature and technology, but producers of energy through our own bodily processes, such as running and riding bicycles and dancing. Most people tend to look at mountains and rivers, so large and forceful, so far beyond human scale, and think, "That's not me. That mass of rock and that flood of water are unrelated to a small, struggling being like me, simply eating, breathing, and walking down the street." Yet our own muscles are mysteriously akin to the flow of rivers and the geological exertions of mountains, when viewed through the lens of energy. The selections in Chapter Two reveal these deep connections. In William Stafford's poem, for instance, the muscular process of biking home from work enables the speaker to perceive the vibrations of meaning and alertness that permeate the world itself. The idea that energy inheres within human nature as well as the outside world appears throughout this part.

Chapter One

Reflections Upon the Grid

Gary Snyder, "Oil" (1958)

b. 1930

Gary Snyder, author of nearly twenty collections of poetry and prose, won a Pulitzer Prize in 1975 for *Turtle Island*. He spent his childhood in Washington State and Oregon, publishing his first poems in a Reed College journal in 1950. Snyder worked seasonal jobs after college, including time as a fire lookout, and spent much of the following two decades studying in Japan. In 1970, Snyder moved to the Sierra Nevada Mountains near Sacramento—or as he might explain in more explicitly bioregional terms, the watershed of the South Yuba River—to build a home which has never been on the electrical grid.

Discussion Generators

Gary Snyder wrote the poem "Oil" in 1958, after a stint working in the engine room on an oil tanker transporting petroleum from the Persian Gulf to U.S. Navy refueling stations in the Pacific Ocean. After this experience, Snyder became convinced—as he explained at a reading of the poem in 2012—that "fossil fuel energy use is like a drug for industrial society." Notice, though, that the first three stanzas focus on the weather, ocean swells, the sleeping crew, and the humming of the ship's furnace—sounds and images that do not directly mention fossil fuels at all. In what ways, then, does this poem comment on industrial society's addiction to fossil fuels? How do the poem's sensory details advance its overall purpose?

soft rainsqualls on the swells
south of the Bonins, late at night. Light

from the empty mess-hall
throws back bulky shadows
of winch and fairlead
over the slanting fantail where I stand.

but for men on watch in the engine room,
the man at the wheel, the lookout in the bow,
the crew sleeps. in cots on deck
or narrow iron bunks down drumming
passageways below.

the ship burns with a furnace heart
steam veins and copper nerves
quivers and slightly twists and always goes—
easy roll of the hull and deep
vibration of the turbines underfoot.

bearing what all these
crazed, hooked nations need:
steel plates and
long injections of pure oil.

Gary Snyder and Kyhl Lyndgaard, "A Human Experience of Energy: A Conversation with Poet Gary Snyder" (2006–09)

Kyhl Lyndgaard, b. 1977

In an e-mail, Snyder noted that "[t]his little interview, when I look at the title, is just a tentative first step toward trying to talk about that 'human experience . . . .'" Originally published in the journal *ecopoetics* in 2009, the interview was conducted at Kitkitdizze, Snyder's home on the San Juan Ridge near Nevada City, California.

Discussion Generators

During a tour of Kitkitdizze, Snyder joked that he had applied to the state of California for a rebate for his wind- and solar-powered clothes dryer, which consisted of a simple clothesline strung up. This playfulness may seem quite different from the somber tone of his poem "Oil." How important is it to keep both good humor and hard realism intact when considering energy and lifestyle choices? What aspects of energy were missed in this "tentative first step" toward a human experience of energy?

GS—What is the original human experience of energy? Extreme cases, like volcanic eruptions. A tornado coming across the plains . . . and think of that little cluster of tipis, watching it coming at them!
KL—The ancient Egyptians depicted electric eels in hieroglyphics.
GS—How strange that must have seemed. Any other examples?
KL—Well, static electricity.
GS—Of course. Other biological systems must use electricity. In some parts of the world, thunder and lightning must have been spectacular. Just spectacular. And what is that all about? Cosmic events, meteorites . . .
KL—Aurora borealis . . .
GS—Seismic energy, the power that is out there and in there . . .
KL—I wanted to ask if you are familiar with William Rueckert's essay "Literature and Ecology"? He's the person who apparently coined the term "ecocriticism."
GS—Oh, is he really?
KL—Seems to be. It's from 1978, and is republished in the anthology on ecocriticism by Cheryll Glotfelty and Harold Fromm.
GS—Tell me more.
KL—Here's a quotation from the introduction: ". . . if poets are suns, then poems are green plants among us for they clearly arrest energy on its path to entropy and in so doing, not only raise matter from lower to higher order, but help to create a self-perpetuating and evolving system."
GS—Oh, ok. That is an interesting little bit of logic, an original image. I'm surprised I haven't read that earlier. I once proposed a similar line of thought in an essay in *The Real Work* which takes a different tack. Instead of speaking of poems as "green life" I propose

them to be mushrooms—as the fruiting bodies of a broad social mycelia. Have you seen that?

KL—It's in *The Real Work*?

GS—At the back of *The Real Work*, in "Poetry, Community, and Climax." And then it's quoted at length in Jed Rasula's book *This Compost*. I'll try to compare Rueckert's image and mine. He proposes a green grazing-cycle metaphor, and I propose a detritus-cycle metaphor of spiritual and artistic energy. I'm looking at the composting of that which is discarded or overlooked. The energy is still usable, in an anti-entropic way.

Not that one approach is more right than the other, they both work toward the same end. What's that quote from Thoreau? Something like the best literature requires the deepest compost. I'm misquoting it, but it's found several places. Try Rasula. He brings it forward and cites it.

KL—Good! I also wanted to ask about the idea of "fossil love" from *Turtle Island* in the poem "As For Poets."

GS—Well, that's a risky one. But here's how it works: the fire poet burns at absolute zero . . . and I start with a simple fact: "At fifty below / Fuel oil won't flow / And propane stays in the tank."

It just hit me as I was working on that poem that "fossil love" might be a term for frozen love, or inaccessible love, or love so deeply buried that it has become rigid, suppressed, hidden. Powerful emotion pushed back into our deeper and darker selves, that sits there as a final remarkable untapped source of energy that might break out sometime. Psychically speaking—like some deeply repressed shame, and when repressed powers come out, they break forth with astonishing force.

The "Fire Poet," whoever he or she might be, is not simply a brilliant hot character manifesting the usual model of heat. "Oh he's hot, she's hot." You know, it's not that easy. I start with the depths of cold to find that the heat of the fire poet might be that which is the most occulted and covered. You might say a little like a volcano . . . icy and cold, but when surface water makes its way down to the magma deep below it all explodes.

Where else did I talk about that? Well, it's in Tibetan psychology also. Deeply buried and suppressed emotions are potentially very powerful and can go both ways. But. I have been told that this is part of the work of some Tibetan/Vajrayana meditations. In some Noh, plays the narrative concerns unresolved transgressions or regrets, and the play itself brings them back to the surface. Dancing and pacing and chanting it through, the protagonist is then liberated from that bit of hard old karma.

The alchemical idea—also common in Tibetan thought—of the transformation of extremes is that poison becomes medicine. That which is most toxic and bitter can be turned 180 degrees into wisdom. The highest wisdom is counterbalanced by deep and bitter angers. Turn it over, get it out.

These are interesting aspects of the work of art and also some religions.

Then there's one of those great essays by Dogen. He considers firewood and ash. Firewood is always firewood, ashes are always ash. Some might think you burn firewood and it becomes ash. But wood also will always be wood, ash will always be just ash. That's another principle—that a thing in its own suchness will always be just that.

KL—So what is it now? [Gestures at the fireplace right in front of us, with burning oak wood.]

GS—Energy! That's the third thing, and Dogen didn't mention it! These are funny little things to work with.

KL—Speaking of firewood now, has it been a useful exercise to have your heat source visible?

GS—Yes, it's like knowing where your food comes from. It's also useful to know the pathways of your firewood. Wes Jackson has a very good essay on this. We talked about this right here where we're sitting. Our firewood here has a five- to five-hundred-year pathway. You might be picking up some wood that died fifty years ago, older than that you wouldn't bother.

The energy in wood is not very old. It comes down as sunlight every year. There is a rapid transformation from that light into green growth, into fiber. It's a short path. Crude oil becomes gasoline and kerosene, and has come on a thirty-five- to forty-million-year path to get here. And then, nuclear energy is as old as the universe. It's as old as matter itself. I don't know why it should be this way, but the longer the path that energy has to travel, the harder and more tricky it is for us to use it.

KL—So, what about your own solar panels? Or what about wind energy?

GS—Yeah! My batteries get charged straight from the sun, or even better, hook up a light directly to the PV panel. Instantaneous transformation! [laughter] Hook up a team of chipmunks to a wheel, set up a waterwheel. I saw a guy with a homemade waterwheel over in the Middle Fork [of the Yuba River]. Really nice, he made it from wood. That was tied into an alternator out of a car, and set in a hand-dug channel for better control. He knew a lot about machinery and got the gear ratios just right. He showed me how his lights came on. He pulled a cord, and the cord opened up the channel and the water ran down the chute. As the wheel picked up speed, the lights got brighter and brighter.

KL—Do you have a compost pile somewhere around your house?

GS—We used to put our wet organic garbage into the forest, but we don't do that anymore, it brings in various critters, especially the bears. They get into everything! Once they hang around, it's hard to get them to go away. The next step, they look into your car, bother you when you want to bring in groceries. We have a well-fenced garden, and a compost bin inside the fence. I never put any bones in the bin, I just burn those with wood trash. That practice really changed the bear numbers. I still see them on occasion, but not nearly as much as I used to.

A little more about energy in this householding life. Take insulation. You don't need to simply harness energy, you need to preserve it. You got a little heat in your body, you save it. Carole [Koda, Snyder's wife] at one point during her long illness was feeling very scattered, was feeling she wasn't getting things done right. She was an ambitious person, in a very good way. She wanted to accomplish things and make a positive difference. So she was trying acupuncture, getting it done by a Chinese person, and she was hoping to feel less scattered in her daily life. The acupuncturist told her she was squandering her *qi*. She was doing things too fast. She let her feelings and opinions get too strong. "Don't do so much, hold on to your *qi*," he said. Conserve your *qi* and don't sweat the small things. That's another kind of energy. *Qi*, an old Chinese term, refers to a sort of non-metabolic spirit energy—is it only metaphorically true? The energy it takes to go through a long Zen *sesshin*—intensive meditative week—is not just energy from food and warm clothes. It's intention, will, a vow, a commitment, to—for—what you don't know.

E. C. Pielou, "Energy Is Everywhere," from *The Energy of Nature* (2001)

b. 1924

In 2001, statistical ecologist Evelyn Christine Pielou published *The Energy of Nature*, which explores energy's role in nature: how and where it originates, what it does, and what becomes of it. Drawing on a wide range of scientific disciplines including physics, chemistry, biology, and all the earth sciences, as well as on her own experiences as a naturalist, Pielou examines the various ways the transfer of energy affects the earth and its inhabitants.

Discussion Generators

Without energy, Pielou reminds us, nothing would ever happen. The sun wouldn't shine, the wind wouldn't blow, seeds wouldn't sprout, birds wouldn't fly, and fish wouldn't swim. In reminding us of the all-encompassing nature of energy, Pielou shows the centrality of energy to our existence: no life can exist without some consumption or transfer of energy. Is this a very simple or a very profound insight? Like the "city child" that Pielou chides for believing milk comes from a carton, what sorts of assumptions about energy might we be guilty of? The scale Pielou's writing takes on is massive, and her writing about numbers in the first half of the selection is remarkable. Examine some of the most engaging writing strategies she uses when grappling with large numbers as she works to make them relatable to a general reader.

In The Beginning

Once upon a time, about 15 billion years ago, the universe—or more cautiously *this* universe—was brought into existence by the Big Bang. At the very first moment, it had zero volume and must have consisted entirely of radiant energy. The density of the energy would have been infinite, and the temperature was of the order of 10^{32}°C. It immediately began to expand and to cool, and it has been doing so ever since.[1] As soon as its volume exceeded zero, things began to happen. By the time the infant universe was 10^{-43} seconds old (that's 0.00 . . . 001 with forty-four zeros), it had grown appreciably, but it was still smaller than a pinhead, about one millimeter in diameter. It was a tiny, expanding fireball, exceedingly dense and intensely hot. A very small fraction of its energy had become matter. From that day to this, energy plus matter has constituted the whole content of the universe.

While the universe aged from five minutes old to about 100,000 years old, it consisted almost entirely of radiant energy plus a plasma of hydrogen nuclei (protons), helium nuclei, and free electrons; no atoms existed until after the end of this stage. Keep in mind that the universe was 100,000 years old about 14,999,900,000 years ago; it had completed a mere 0.00007 of its life span to date and was still in its infancy. By the end of this stage the temperature had dropped to about 100,000°C, making it possible for protons and electron to combine in pairs and create hydrogen atoms. (That the 100,000-year-old universe had a temperature of 100,000°C is simply a coincidence; the numbers are only approximate in any case.) Thereafter galaxies and stars started to form, and the energy in matter began to exceed the energy in electromagnetic radiation. The changes have been continuing in the same direction ever since—less and less radiant energy and more and more energy in matter—and will no doubt go on doing so. That is, conditions in the universe haven't changed qualitatively for the past 14,999,900,000 years; but its temperature continues to drop, it continues to expand, and matter becomes increasingly dominant relative to radiant energy.

1. Arthur Beiser, *Concepts of Modern Physics*, 5th ed. (New York: McGraw-Hill, 1995).

After that brief account of the history of the cosmos, let us return to life on our planet. It is a world where things happen, and happenings always entail energy. Even the moon is not truly a dead world, in spite of its bad press. It may lack life in the usual sense of the word, but things happen there: meteorites strike it; the surface heats under the sunshine and cools during darkness, making the rocks alternately expand and contract so that they fracture; the fragments fall. And whenever anything is happening, energy is being transferred from one piece of matter to another.

It surely follows that energy should attract the attention of observers at least as strongly as "things" do. Everybody is surrounded all the time by energy transfers: events, actions, "happenings." It's worthwhile to consider the implications, especially for naturalists.

A Hike in the Country

Imagine a hike in the country and the things an observant hiker would see. The list will probably include many living things: trees, flowers, birds, butterflies, perhaps squirrels and deer. There will also be scenery: rivers and streams, lakes, ponds and marshes, mountains and hills, perhaps beaches and the sea, and for skywatchers, blue sky and clouds by day or the moon, the stars, and maybe (with luck) a comet by night. The list can be extended almost indefinitely. It is a list of *things*, however—material things—and it represents no more than half of what surrounds the hiker. The scene is also filled with energy: not directly visible, it is true, but rendered observable through countless actions, movements, and events.

Imagine the scene once more, this time concentrating on all the signs of energy to be seen: twigs and branches swaying in the wind, scudding clouds, flowing water, breaking waves, flying birds and insects, running deer. Things both living and nonliving are continually moving, a sure sign that energy is being spent. Think of the sounds the hiker hears, for sound is a form of energy: the crackle of dry leaves underfoot or the drumming of rain, the babbling of a stream, the calls of birds, the hum of insects. Sound is much more noticeable on a windy day, with the roar of wind and waves at the beach and the snapping of tree branches in the forest. The stormier the weather, the more obvious the energy. Lightning gives a glimpse of yet another of energy's many forms—electrical energy.

Movements, sounds, and the occasional lightning flash are merely the more attention-getting forms of energy. The warmth and brightness of sunshine and the growth of plants illustrate how the sun's energy empowers life and action at the surface of the earth; energy from the sun comes as electromagnetic radiation, and plants grow because they can convert the radiant energy into chemical energy.

Energy in a multitude of forms is as much a part of our surroundings as are tangible things, and it is just as noticeable to anybody who pays attention. In the city, evidence of energy at work—man-made energy—is impossible to avoid: think of the roar of traffic, the bright lights, the construction sites with cranes and concrete mixers, even the din of shopping-mall music. But energy is as abundant in the tranquil countryside as it is in the city, since all energy has its ultimate origin in natural sources exactly as material substances do. Imagining otherwise is like a city child's not believing that milk comes from cows because it so obviously comes from cartons.

Energy is as much a part of nature as matter, and all artificial energy derives from natural energy. Coal, oil, and natural gas are stores of fossil solar energy. Hydroelectric power is simply solar energy that has been converted to human use more quickly. Nuclear energy existed as natural energy for billions of years before humans built nuclear power plants. Knowledge about energy is knowledge about the basic workings of the universe and is fundamental to all of science; it is not simply part of engineering. Name any branch of science—physics, chemistry, biology, geophysics, oceanography, meteorology, quantum mechanics—and you will find it is about energy as much as about matter. From black holes and supernovas to viruses and genes, "things" of all kinds have both energy and matter; their energy is as important a part of them as their matter.

Vachel Lindsay, "A Rhyme About an Electrical Advertising Sign" (1914)

1879–1931

Vachel Lindsay is considered to be the progenitor of modern singing poetry, as he referred to it, in which verses are intended to be sung or chanted. His correspondence with Irish poet William Butler Yeats outlines his intentions to revive the musical qualities in poetry that had been celebrated by the ancient Greeks, among others. Lindsay's furious, fiery, and often zany approach is evident in the following poem, "A Rhyme About an Electrical Advertising Sign."

Discussion Generators

In this poem, Lindsay grapples with the technological advances that made possible the electric advertising sign, then a simple electric discharge lamp consisting of a transparent container in which a trapped gas was energized by an applied voltage. Although the neon sign had not yet been patented—that would happen the year after the poem was first published—Lindsay anticipates the ways that electrical advertisements would transform the way people perceive the night. What language still rings true today? What new language might be used to capture the way the night sky worldwide has been altered by the ubiquitous presence of artificial light?

I look on the specious electrical light
Blatant, mechanical, crawling and white,
Wickedly red or malignantly green
Like the beads of a young Senegambian queen.
Showing, while millions of souls hurry on,
The virtues of collars, from sunset till dawn,
By dart or by tumble of whirl within whirl,
Starting new fads for the shame-weary girl,
By maggoty motions in sickening line
Proclaiming a hat or a soup or a wine,
While there far above the steep cliffs of the street
The stars sing a message elusive and sweet.

Now man cannot rest in his pleasure and toil
His clumsy contraptions of coil upon coil
Till the thing he invents, in its use and its range,
Leads on to the marvellous CHANGE BEYOND CHANGE.
Some day this old Broadway shall climb to the skies,
As a ribbon of cloud on a soul-wind shall rise.

And we shall be lifted, rejoicing by night,
Till we join with the planets who choir their delight.
The signs in the street and the signs in the skies
Shall make a new Zodiac, guiding the wise,
And Broadway make one with that marvellous stair
That is climbed by the rainbow-clad spirits of prayer.

David Gessner, "Energy," from *The Tarball Chronicles: A Journey Beyond the Oiled Pelican and Into the Heart of the Gulf Oil Spill* (2012)

b. 1961

The author of nine books of nonfiction, David Gessner teaches at the University of North Carolina at Wilmington. Gessner founded the literary journal *Ecotone*, and his work has garnered awards ranging from a Pushcart Prize to the John Burroughs Award for Best Natural History Essay. His book *The Tarball Chronicles: A Journey Beyond the Oiled Pelican and Into the Heart of the Gulf Oil Spill* grew out of a series of blog posts written for the journal *OnEarth*.

Discussion Generators

Gessner's writing has long been noted for its humor, yet *The Tarball Chronicles* looks at the unlaughable Deepwater Horizon oil spill (also known as the BP oil spill) that began in April 2010. Gessner's website notes that this book is a "new way of writing about nature and place, full of humor and strangeness, stripped of the old pastoral cliches, and focused on a more 'limited' nature, the only nature left to most of us." Upon reading the excerpt below, do you think Gessner has succeeded in his task? Gessner also looks beyond typical energy discussions to emphasize the primal relationship between human energy, creativity, and the natural world. How would you harness our natural tendencies toward movement and creation in a positive way?

Lofty thoughts give way to base needs. We are tired and we are hungry. When we get off the water we decide to have lunch at a tiny Vietnamese restaurant called Pho. There is a substantial Vietnamese fishing community in Bayou La Batre, and this is where they come for noodles. The Vietnamese owners have seen a lot in their day but I am confident that they have never seen anyone put as much hot sauce on their noodles as Hones does now. When the broth looks bloody enough, he goes to work slicing up little peppers and plopping them in. Hones is a pepper freak and during the trips he's often fretted about the pepper plants he left untended in his apartment back home.

When I don't finish my beef and broccoli, he takes it and drowns it in hot sauce. When Bethany doesn't finish hers, he asks if he can polish that off, too.

"I don't like wasting food," he says by way of explanation.

"You sure don't," I say.

He ignores me and reaches for Bethany's bowl.

"I'll give it a good home," he promises, rubbing his belly.

After lunch we thank Jeff and hug Bethany good-bye. She is going back to Mobile and we are heading on to Louisiana. I nap through much of Mississippi while Hones drives. He gets us past New Orleans and then, on the way down to Buras, we stop for beer and gas and I take the wheel. We pass signs that say "Plaquemines Parish Thanks Billy Nungesser" and billboards that ask "Oil Spill Claim?" and advertise for lawyers. This is Hones's first time here, and as we descend into the sunken land he takes it all in. He points out birds and another sign that reads "Elevation 4 Feet." I'm actually surprised that those numbers are positive. If most of the world's scientists are right, this will all be underwater by the time my daughter is my age. But no one quite believes it—people *live* here, after all.

It's dark by the time we reach Buras. We head, not to Ryan's lodge, but to the environmental organization's headquarters, where we will spend the next couple of days. Before we came down the organization offered to pay Hones for his photographs and videos, which I thought a coup for a guy who had recently lost his job. But he declined, worried that it might somehow disturb his unemployment checks. He did agree to sign a release, however, so we now have a place to sleep. I am particularly happy with the fact that we can take bunks, and rooms, on opposite ends of the hallway, effectively ending the snoring crisis.

After we've thrown our things in our rooms, we head over to the lodge, where Ryan invites us to his house and announces that "something is not right" out on the water. We listen for a long time to his grim message. More happily he also tells us, as we are leaving, about a place to fish near the lodge. It's late by the time we get back to the headquarters. We drink beers and turn on the TV and watch some Time Life infomercial about singer songwriters from the seventies. In between snippets of John Denver and Cat Stevens, we talk about Ryan. For his part, Hones can't believe he was in the same room with the guy who lives out his own dream job. It took about a minute for him to be completely in Ryan's thrall.

"That guy's fucking amazing. He really *knows* his place."

A high compliment in Hones's book.

We sleep well and the next morning we drive down to Venice. Not long after Halliburton Road, I pull over and take out my binoculars. Just like this summer, there are plenty of egrets and herons and cormorants. A black rail runs across the road, a sneaky dark-cloaked spy caught momentarily in the open. Hones can see fish jumping and is tempted to take out his rod, but decides to wait until the afternoon. I notice that there are many more ospreys than there were this summer. For some this must be their final migratory destination. To the south, out by open water, I see some large birds that I think might be gannets, but then again that might be wishful thinking.

What we do see, right before we leave, are a half dozen wheeling white birds that gradually clarify themselves into pelicans. For the second time in two days I am being treated to the sight of white pelicans, this group of five soaring upward in great swooping spirals.

Their grace is impressive, but so is their sheer size. To put this in perspective, consider than an osprey, a bird that many nonbirders confuse with eagles, usually weighs about four pounds. People can't believe it when you tell them, but ospreys, like all birds, weigh next to nothing due to their hollow bones. Then consider that brown pelicans, generally considered a *huge* bird, weigh twelve pounds. Finally, with that as context, turn to the white pelicans that we now watch as they, seemingly weightless, spiral upward. Their weight? Twenty-five pounds.

I remember something that the guy in the bar in Mobile said.

"Brown pelicans are tough birds. But one white pelican can kick the asses of two or three brown ones."

The birds vanish, becoming white specks, then nothing. We climb back into the car and drive north to Buras. Ryan had mentioned that he might take us out on the water, but when we stop by the lodge he says it's too windy. The weather has shifted dramatically. This morning we were sweating, Hones especially, but around noon a cold front swooped in and the wind came down from the north. The fishermen staying at Ryan's lodge are disappointed, but Hones is relieved. He hates the heat. In the late afternoon we pull into one of the few restaurants in Buras, a drive-through daiquiri bar. That it's a strange operation goes without saying, but the bartender tells us that daiquiri bars are common in Louisiana. We sit on stools in the bar, while over in the corner men and women play video slot machines. I order a margarita and Hones a piña colada, and we sip them while the bartender, who grew up here, tells us how beautiful and bustling the town was before Katrina.

We rest briefly back at the headquarters and then I drive Hones to a fishing spot that Ryan told us about. It is off Route 23—*everything* here is off 23—over the canal where we saw an alligator earlier and up a dirt path to the hump of land on the Gulf side. I park on the dirt and we wander toward the tall grasses and reeds and, sure enough, there's the gap in the grass just where Ryan told us we'd find it. The sunlight is fading as we cut into the path and I'm a few feet ahead of Hones. I step over a log and am about to put my foot down when I notice it. Big and black, yellow underneath, semicoiled. It looks as thick as a python, and I'm pretty sure it's the biggest snake I've ever seen, including the boa constrictor Hones and I once saw in the Belizean rainforest. My foot freezes in the air, and the snake pauses, as if casually deciding whether to kill me or not, before slithering off into the brush. I jump back jittery.

"What the hell happened?" Hones asks.

I tell him what it looked like and we are pretty sure it was a cottonmouth, aka a water moccasin. I have seen plenty of snakes but for some reason this one leaves me feeling shaky. For all I love nature, I'm also fond of life.

"I've never seen you like this," says Hones.

My normal bluster has deserted me. I am blusterless.

"That thing was big," I say. "And scary."

When we head down the path again we walk more carefully. Hones finds a good spot and sets up a little camp for himself and I promise to be back to pick him up in a couple of hours. When I walk out I take extra care stepping over the log.

After my encounter with the snake, I drop by the lodge to visit Ryan. He asks where Hones is.

"At the spot," I say.

Ryan laughs out loud.

"The spot!"

He finds the name ridiculous, even though he was the one who told Hones about it. For Ryan any sort of land fishing is junk fishing, not comparable to what you can reach by boat. But for Hones it works out just fine. When I left him he was overjoyed with his new cottonmouth fishing spot and over the next day and a half he will catch a variety of fish both there and down in Venice—catfish, trout, redfish, drum.

Since Hones lost his job, he's been heading out to fish at Wachusett Reservoir every day, and he was reluctant to come on this trip lest he miss some of the last days of fishing season. The spot is a fine substitute: his own private Wachusett.

For my part, I'm glad to get some time alone with Ryan. Who knows when I'll be back? We sit at the long dining room table and tease each other about the coming election. That we can talk about it at all makes me feel better about the country.

"When we can only think in opposites, conservative and liberal, black and white, we can never reach any creative solutions," he says. "We have the same goals in mind, Professor, just a different route."

He tells me more about his childhood. He rues his lack of education. Straight to work after high school. Doing time at the chemical plant for twenty-one years. But still getting down here to fish and guide. "I was always driven, for some reason," he says. "Guiding was my part-time job. Only it was a full-time part-time job.

"I taught myself Spanish and I read a lot. And I've learned about nature and what makes her tick. And she's taught me. I know more about nature than most scientists, though they know the inside workings better than me."

Ryan, I understand, is not perfect. "Don't bullshit a bullshitter," my father used to say. Ryan is putting on a show for me, the scribe. He is a big man with a big personality and a big story to tell and he is not going to apologize for that. I have dined with him and a group of fishermen while he entertained them with fish stories—violent, aggressive, sometimes bawdy stories that always starred Ryan Lambert—and remember that he mentioned offending someone and then added: "What are they going to do? They can't kick my ass and they can't get me thrown into jail for talking." I can see that the conversational mode he is most comfortable in is the monologue and I am fine with that.

But there are other parts, too, parts that I think I understand. Like him, I don't have much use for sleep. Like him, I had a father who died young. Like him, I am driven.

Maybe what I admire most about Ryan is the way he rebuilt after Katrina. It is a capacity I know I have in myself. I hope I don't say that boastfully. It's just that I have come to learn that the way I respond to devastation—personal or professional—is with a redoubling of effort and an angry determination to begin again. "I was possessed," Ryan said to me about rebuilding his lodge after Katrina. Exactly.

When Orrin Pilkey first told me that he thought the rebuilding of beach houses after storms was "societal madness," I nodded my head in agreement. I was with him in spirit. I am still with him on most things, but I can no longer nod about that one. If it is a madness, it also neighbors one of our best human qualities. A determination, a fury, a fuel to make something, not just out of oneself, but out of the world. It is how things really get done on this planet.

We are sitting on the living room couch when I tell Ryan how impressed I am with his speechifying. When I listened to the tapes I'd recorded during the summer what I noticed most, other than his passion for this place, was the fact that there were almost no "uhs" or "ums" during the course of the interviews.

He seems pleased by this. He admits that he has considered running for office on a "Free the Mississippi" platform. I jokingly suggest that we should run for national office together, uniting the Red and Blue, and he agrees, laughing. But then I think about Ryan adding the burden of public office to the burden of all he does already, and I ask seriously: "Do you have the energy for that?"

As soon as it escapes my lips I know it is a stupid question. Here is a man who worked nights in a chemical plant and days as a bayou guide. Here is a man who has come back from utter devastation, his dreams shattered and flooded. Here is a man who rebuilt those dreams, and his lodge, after Katrina, along with rebuilding a significant portion of Plaquemines Parish.

"Do I have the energy?" he repeats.

He looks at me dumbfounded.

"I *am* energy," he says.

On the last morning of my gannet trip to Nova Scotia I got up before dawn. I wanted to witness a spectacle of raw energy called the tidal bore. Within the course of half an hour all the waters near the Bay of Fundy begin a mad rush down dozens of tidal creeks, coming in powerful waves, and those waters fill the empty bay like a giant bathtub. It was summer, when the gravitational pull of the moon, and the sun, is strongest, making the tides all the more dramatic. In the middle of the night—at 3:30—I joined a ghostly procession of a dozen other guests streaming from their hotel rooms toward the water. We stared out at the empty mud chasm, about a hundred yards wide, where the Salmon River ran when it was in the mood. As if on cue came a noise like a train rumbling, the rumbling quickly growing louder. Then a wave, breaking in slow motion, charged toward us, filling the dry basin. While the wave wasn't huge—nothing you could surf, though I'd heard people try—the phenomenon, like that of a flash flood, was full of power and suddenness. Within ten minutes a river ran at our feet where a river hadn't been, gulls hitching a ride on the current while swallows darted and zipped, following the insects that followed the water. Once the basin filled, my fellow onlookers left, and I stayed alone as the river spilled over the brim. Then, in an exciting if not climactic finale, a kind of denouement, the river started to flow and burble in the opposite direction, the water reversing itself against the incoming tide. There was no one moment when you can say the river was really still. No moment either when you could clearly see that a switch had been thrown. But suddenly, on some parts of the water eddies formed and on other parts where sticks had been floating one way, they now floated another.

I felt electrified, and not just by the thrill of gushing water. Think of it. Approximately 100 billion tons of water flow in and out of the Bay of Fundy with every tide, the equivalent, scientists say, of the combined flow of all the freshwater rivers in the world (though one wonders how this could possibly be measured). As it turns out, the place has understandably inspired its own Jim Gordons and tidal energy plants are in the works that could produce 100,000 megawatts of energy, enough to power 100,000 homes.

Wind, tides, sun. We need to use it all. And while we're at it there is another source that we had better not forget about. As we search high and low for different types of energy, we had better not forget the human kind. How do we use that human raw material? This is a question every bit as pressing as "Will ethanol work?" How do we free up the sort of energy that Ryan has and let it work for, not against, the world? How do we tap it? How do we created *driven* environmentalists?

I'm not sure. But it doesn't seem to me that we can do it softly at this point. It seems to me that the reason so many people admire Teddy Roosevelt, the reason I admire Ryan Lambert, is that they fight fire with fire. They are not anti-energy. They do not think they can force a river to run in straight lines. But they can nudge it, encourage it to move back into older paths. Human beings naturally want to be successful—and the urge to be successful provides the fuel of self-interest—but we must begin by redefining our meanings. Success can't just mean the biggest and most and straightest. Why couldn't you be the most ambitious man in the world and have your ambition be to see how much you can make of little?

It is vital to unite *excitement* with conservation. If we perceive "conserving" as dour, we will all turn the other way. We don't need saints and we don't need hushed quiet. We need Jim Gordon saying "Yes, I want to make money," and we need Jeff proud of piling oysters behind barrier islands, and we need Ryan talking excitedly about freeing the Mississippi. We need not just a hundred miles of oyster reefs, but a thousand. We need Orrin Pilkey taking glee in telling the truth about our barrier islands, despite the wrath of the locals. We need someone with sense, and a sense of nature, to take over the Corps of Engineers and show them, excitedly and once and for all, that straight isn't the only way to draw lines. We need young people excited about making something other than, or along with, money, who still believe in worlds other than the virtual.

The next evening, our last, I drop Hones at his spot for a final night of fishing and head over to my own spot on the Mississippi. I park at the little boat launch on the river, the same one I visited this summer. The sun is dying back by Hones, over the millions of acres of wetlands, and that death is reflected in the sky over the river in pinks and blues. An osprey flies along the river's banks, riding the north wind that is blowing down clear from Minnesota and Wisconsin. The same wind brings the great final sweep of late migration, including loons, from the north.

I stare out at the river and I am happy, stupidly happy. We are on the brink of one of the most divisive and acrimonious elections in our history—so what's new?—but for the moment that particular peep show, that carnival of charlatans, doesn't matter. Neither does the fact that oil is still washing up on shore or that the gills of the shrimp here are black as tar. No, what matters right now is the roseate spoonbill that flies overhead, its feathers almost exactly the same pink as the gaudy line of light below the blue clouds.

This show is here every night if you want to come see it. Wondrous things are always happening. I think of the people who first migrated to this land, whether by ice bridge or ship, and what they found: a land of wildly extravagant resources and wildly extravagant beauty. What we've done to that land and those resources would almost be comic if not for the tragedy, and we've done it with the alacrity of some crazy, sped-up film. Can we possibly redeem ourselves? *Fat chance*, says my cynical side. *Please, oh please*, says the other. By now I know that when Ryan talks about loss he is not just addressing the loss of species and wildlife habitat. He is talking about the loss of a way of life. Not just hunting and fishing either, but the loss of a life connected in any way to the natural world. And as fewer and fewer of us are connected, fewer of us understand what we are losing. It's not about environmentalism. It's about this beautiful gray river and the osprey and the line of clouds and the roseate spoonbill. It's about what is best even if we sometimes forget what best really is. It's about wildness, a wildness that is still there inside our human chests and that vibrates like a tuning fork when we see a match for our wildness in the world.

The last streaks of sun blaze a gaudy pink. The water takes on a blue-black metallic sheen. Wind whips the sedge grass. Do we really want to destroy this? Do we really want to cut off the connection to the animals we are?

Maybe the trick, if you can call such a profound thing a trick, will be in redefining what we mean by self-interest. Environmentalists like to argue that saving an ecosystem is "useful," and, given their opponents' attitudes, this is perhaps a sensible approach. But it's such a limited way to think. By agreeing that that's the table the game will be played on, we tilt the game itself. Rather than *use*, it is the sheer wild uselessness of nature, the sheer nonutilitarian, unrelated-to-human uselessness, that is cause for celebration. I have always thought that nature was the source of my creativity, and the source of creativity for most artists, even those who have never set foot on a beach or in the woods. But my thinking is evolving and I am moving beyond those inchoate ideas. I am coming to believe that nature *is* creativity. Not just a wellspring for humans but the thing itself.

Isn't it in our self-interest to hang on to that wellspring? Or at least to make sure that a path back to it exists? How odd that we are destroying the very thing that lets us imagine more. And by extinguishing things at the rate we are—there's no need here to drag out the well-known roll call of extinction and habitat destruction—we are doing no less than putting an end to creation. This may sound like overstatement. It isn't. For millions of years different species have evolved from each other in a thousand miraculous ways—*shrews becoming dolphins!*—and now we have said, "Enough is enough. Time for this messiness to end." We are busy neatening up the world, getting it organized, making it useful for humans, building our castles.

At my most pessimistic, I see no way of halting this. After all, we can never stop banging on things, can never stop making and improving. Most of us are no more capable of keeping still than beavers, whose own teeth impale them unless they constantly gnaw. So how can we, busy beavers that we are, have any hope of not chewing down all our trees and damming the world?

We are told that the castle builders and straight-line thinkers are the ones who understand the "real world." We'll see. The proof is in the pudding and if this summer is any indication, that pudding will be served up soon. We are also told that there is no time to get sentimental about nature, and anyway nature is in the way—it simply must go. And the scary thing is that most of us nod numbly and agree, if not in theory then in practice. But with every step forward we lose the path back. The more the straight-line thinkers win out, the less of a source is left for creativity. The less creativity, the less chance of getting back. When we kill the woods or beach we are killing possibilities. Our options, biologically as well as artistically, become limited. After all, you can't simply re-create dolphin or pelican or kangaroo.

I could go on but I will stop my preaching now. I am tired, weary. One of the things that straight-line thinkers like to do is to segregate, keeping everyone and everything in their separate cells. In this way, we can focus on the narcotics of our specialties: macramé or biochemistry or golf. In my field this means keeping art separate from politics, which is one of the rules of literature in the past century. It is a rule that I would, quite honestly, like to follow and one that I did follow for the first twenty years of my career. But it just doesn't seem possible anymore, what with the world ending and all. I'd like nothing more than to hole up in my garret and make art. But I can't. I am leaving this place tomorrow. But I will remain part of this mess forever.

William Rueckert, "Literature and Ecology: An Experiment in Ecocriticism" (1978)

1926–2006

In his 1978 essay "Literature and Ecology: An Experiment in Ecocriticism," first published in the *Iowa Review*, William Rueckert may have been the first scholar to use the word "ecocriticism," a term he defines as "the application of ecology and ecological concepts to the study of literature." In the decades since its original publication, Rueckert's essay has been a crucial influence on the growing field of ecocriticism, an important movement in literary and cultural studies. Among other things, ecocritics study the human perception of nature, how it has changed throughout history, and whether current environmental issues are accurately represented or even mentioned in literature and culture.

Discussion Generators

Rueckert posits that literature—what he calls "poetry"—must be understood as a form of potential energy which, when activated by the reader, transforms into kinetic energy. Literary production is then inseparable from natural forces. List the many ways that Rueckert considers literature to be a form of energy, and find the critical ways he believes human energy differs from natural energy. Do you find these arguments to be persuasive? Which are most problematic for you? What are some of your own cultural touchstones—literature, music, film, etc.—that you return to again and again?

I'm going to begin with some ecological concepts taken from a great variety of sources more or less randomly arranged and somewhat poetically commented upon.

A poem is stored energy, a formal turbulence, a living thing, a swirl in the flow.

Poems are part of the energy pathways which sustain life.

Poems are a verbal equivalent of fossil fuel (stored energy), but they are a renewable source of energy, coming, as they do, from those ever generative twin matrices, language and imagination.

Some poems—say *King Lear*, *Moby Dick*, *Song of Myself*—seem to be, in themselves, ever-living, inexhaustible sources of stored energy, whose relevance does not derive solely from their meaning, but from their capacity to remain active in any language and to go on with the work of energy transfer, to continue to function as an energy pathway that sustains life and the human community. Unlike fossil fuels, they cannot be used up. The more one thinks about this, the more one realizes that here one encounters a great mystery; here is a radical differential between the ways in which the human world and the natural world sustain life and communities.

Reading, teaching, and critical discourse all release the energy and power stored in poetry so that it may flow through the human community; all energy in nature comes, ultimately, from the sun, and life in the biosphere depends upon a continuous flow of sunlight. In nature, this solar "energy is used once by a given organism or population; some of it is stored and the rest is converted into heat, and is soon lost" from a given ecosystem. The "one-way flow of energy" is a universal phenomenon of nature, where, according to the laws of thermodynamics, energy is never created or destroyed: it is only transformed, degraded, or dispersed, flowing always from a concentrated form into a dispersed (entropic) form. One of the basic formulations of ecology is that there is a one-way flow of energy through a system but that materials circulate or are recycled and can be used over and over. Now, without oversimplifying these enormously complex matters, it would seem that once one moves out of the purely biological community and into the human community, where language and symbol-systems are present, things are not quite the same with regard to energy. The matter is so complex one hesitates to take it on, but one must begin, even hypothetically, somewhere, and try to avoid victimage or neutralization by simple-minded analogical thinking. In literature, all energy comes from the creative imagination. It does not come from language, because language is only one (among many) vehicles for the storing of creative energy. A painting and a symphony are also stored energy. And clearly, this stored energy is not just used once, converted, and lost from the human community. It is perhaps true that the life of the human community depends upon the continuous flow of creative energy (in all its forms) from the creative imagination and intelligence, and that this flow could be considered the sun upon which life in the human community depends; but it is not true that energy stored in a poem—*Song of Myself*—is used once, converted, and then lost from the ecosystem. It is used over and over again as a renewable resource by the same individual. Unlike nature, which has a single ultimate source of energy, the human community would seem to have many suns, resources, renewable and otherwise, to out-sun the sun itself. Literature in general and individual works in particular are one among many human suns. We need to discover ways of using this renewable energy-source to keep that other ultimate energy-source (upon which all life in the natural biosphere, and human communities, including human life, depends) flowing into the biosphere. We need to make some connections between literature and the sun, between teaching literature and the health of the biosphere.

Energy flows from the poet's language centers and creative imagination into the poem and thence, from the poem (which converts and stores this energy) into the reader. Reading is clearly an energy transfer as the energy stored in the poem is released and flows back into the language centers and creative imaginations of the readers. Various human hungers, including word hunger, are satisfied by this energy flow along this particular energy pathway. The concept of a poem as stored energy (as active, alive, and generative, rather than as inert, as a kind of corpse upon which one performs an autopsy, or as an art object one takes possession of, or as an antagonist—a knot of meanings—one must overcome) frees one from a variety of critical tyrannies, most notably, perhaps, that of pure hermeneutics, the transformation of this stored creative energy directly into a set of coherent meanings. What a poem is saying is probably always less important than what it is doing and how—in the deep sense—it coheres. Properly understood, poems can be studied as models for energy flow, community building, and ecosystems. The first Law of Ecology—that everything is connected to everything else—applies to poems as well as to nature. The concept of the interactive field was operative in nature, ecology, and poetry long before it ever appeared in criticism.

Reading, teaching, and critical discourse are enactments of the poem which release the stored energy so that it can flow into the reader—sometimes with such intensity that one is conscious of an actual inflow; or, if it is in the classroom, one becomes conscious of the extent to which this one source of stored energy is flowing around through a community, and of how "feedback," negative or positive, is working.

Kenneth Burke was right—as usual—to argue that drama should be our model or paradigm for literature because a drama, enacted upon the stage, before a live audience, releases its energy into the human community assembled in the theater and raises all the energy levels. Burke did not want us to treat novels and poems as plays; he wanted us to become aware of what they were doing as creative verbal actions in the human community. He was one of our first critical ecologists.

Coming together in the classroom, in the lecture hall, in the seminar room (anywhere, really) to discuss or read or study literature, is to gather energy centers around a matrix of stored poetic/verbal energy. In some ways, this is the true interactive field because the energy flow is not just a two-way flow from poem to person as it would be in reading; the flow is along many energy pathways from poem to person, from person to person. The process is triangulated, quadrangulated, multiangulated; and there is, ideally, a raising of the energy levels which makes it possible for the highest motives of literature to accomplish themselves. These motives are not pleasure and truth, but creativity and community.

Simon J. Ortiz, "Electric Lines" and "Gas Lines" (1992)

b. 1941

The identities of poet Simon J. Ortiz are multiple: he has an American name and an Acoma name, Hihdruutsi; he is from Acoma Pueblo in New Mexico, but has lived in various places, even as far away as Toronto, Canada; and he has worked in a variety of jobs ranging from mining laborer to college professor. Drawing on American Indian storytelling, his poems emphasize orality, narrative, and the effects of language on the human and natural worlds. As Ortiz explains, his poetic vision is about more than just style: "The purpose of that story sharing or storytelling is . . . conversing, and the story listeners are conversing with us. We are sharing, or participating. And it's the storyteller participating by his telling, and the listener participating by his or her listening. So it's an exchange. It's a dialogue. It's an event."

Discussion Generators

In the two short poems, Ortiz attempts to make sense of the profound changes that came to Acoma country when electric and gas lines "connected up Aacqu / with America"—changes about which Ortiz is clearly ambivalent. As he notes, his poems are meant to be a conversation between the storyteller and the listener. What is the nature of conversation in these poems? Why would the speaker tell his siblings "some lie" at the conclusion of the poem "Electric Lines," and what is the nature of dialogue between the El Paso Natural Gas Company and the residents in "Gas Lines"? Finally, consider multiple meanings of the word "lines" in these two poems and think about the relationship between our interpersonal connections and our utility connections. What is lost, according to Ortiz, when native communities become connected to the energy grid?

Electric Lines
At first, the electric lines ran
only alongside the railroad tracks
but later they connected up the homes
in Deetseyamah and Deechuna with the lines.
Those electric lines connect up Aacqu
with America.

When they were putting up the lines,
there was this machine.
The machine had a long shiny drill
which it pointed at the ground
and drove it turning into the earth
and almost suddenly there was a hole
in the ground and the machine
drove on to another spot.
A couple of days later, a truck came
and threw long black poles
into the holes. At that age,
I didn't know much of anything,
and when my younger brothers and sisters
asked me what was going on
I probably told them some lie.

Gas Lines
The El Paso Natural Gas pipeline
blew up in the spring of 1966.
Old man Tomato told us what he was doing
on that early morning. He had just gone
out to piss outside his home
and had just come back inside
and was lying on his bed.

"I was singing a hunting song,
and then all of a sudden there was
this strange feeling and then I looked
out the window to the east and saw a light
over the hill beyond Dahska, but I knew
that it wasn't about to be dawn yet."
He told us that at the meeting in Acomita
where the El Paso Natural Gas man said
his company was sorry and they would
send money to make restitution very soon.

The El Paso Natural Gas company ran
through our best garden and left stones.
I was at Indian Boarding School at the time
and so I didn't see it coming through.
It wouldn't have made any difference
whether or not I'd seen it coming through
or whether we'd put a garden in that spring
because it would have come through anyway.
Nobody and nothing could stop it coming through.

Chapter Two

"A Going Thing": Machines, Mountains, and Muscles

Robinson Jeffers, "O Lovely Rock" (1938)

1887–1962

In the 1920s and 1930s, at the height of his popularity, poet, essayist, and dramatist Robinson Jeffers was famous for being a rugged outdoorsman, living in relative solitude and writing of the difficulty and beauty of the wild, especially in the Big Sur region of the central California coast. He spent much of his life in Carmel, California, in a granite house that he mostly built himself. To build the first part of Tor House—a small, two-story cottage—Jeffers hired a local builder from whom he learned the art of stonemasonry. Jeffers's poems, collected in such volumes as *Give Your Heart to the Hawks and Other Poems* (1933) and *The Double Axe* (1948), often reflect his fascination with geology. In his poem "O Lovely Rock," Jeffers approaches the world with loving attention rather than as a riddle to be solved, encouraging readers to see the rock apart from the mediating influences of culture. Jeffers's sense of wonder enables him to experience what he calls the rock's "intense reality."

Discussion Generators

What evidence is there, early in this poem, that Jeffers may be building up to a radical reconsideration of his relationship to stone? He refers to "slide-rock" and "gravel," but aren't these mineral references merely part of the backdrop of the poem's hiking and camping narrative? Something strange happens at night by the light of the small campfire, sparking the poet's "eyes and mind" as he gazes at the rock wall near where his son and his son's friend are sleeping. What insights into the energy of stone, of mountains, does the poet suggest here, and why does this understanding result in "love and wonder"?

> We stayed the night in the pathless gorge of Ventana Creek, up the east fork.
> The rock walls and the mountain ridges hung forest on forest above our heads, maple and redwood,
> Laurel, oak, madrone, up to the high and slender Santa Lucian firs that stare up the cataracts
> Of slide-rock to the star-color precipices.
>
> We lay on gravel and kept a little camp-fire for warmth.
> Past midnight only two or three coals glowed red in the cooling darkness; I laid a clutch of dead bay-leaves
> On the ember ends and felted dry sticks across them and lay down again. The revived flame
> Lighted my sleeping son's face and his companion's, and the vertical face of the great gorge-wall
> Across the stream. Light leaves overhead danced in the fire's breath, tree-trunks were seen: it was the rock wall
> That fascinated my eyes and mind. Nothing strange: light-gray diorite with two or three slanting seams in it,
> Smooth-polished by the endless attrition of slides and floods; no fern nor lichen, pure naked rock . . . as if I were
> Seeing rock for the first time. As if I were seeing through the flame-lit surface into the real and bodily
> And living rock. Nothing strange . . . I cannot
> Tell you how strange: the silent passion, the deep nobility and childlike loveliness: this fate going on
> Outside our fates. It is here in the mountain like a grave smiling child. I shall die, and my boys
> Will live and die, our world will go on through its rapid agonies of change and discovery; this age will die,
> And wolves have howled in the snow around a new Bethlehem: this rock will be here, grave, earnest, not passive: the energies

That are its atoms will still be bearing the whole mountain above:
    and I, many packed centuries ago,
Felt its intense reality with love and wonder, this lonely rock.

Marilyn Nelson, "Minor Miracle" (1997)

b. 1946

Brought up first on one military base and then another, Marilyn Nelson started writing poetry while still in elementary school. She earned her B.A. from the University of California, Davis, and holds graduate degrees from the University of Pennsylvania (M.A., 1970) and the University of Minnesota (Ph.D., 1979). Her honors include two Pushcart Prizes, two creative writing fellowships from the National Endowment for the Arts, a Fulbright Teaching Fellowship, and she served a five-year term as poet laureate for the state of Connecticut. She is professor emerita of English at the University of Connecticut.

Discussion Generators

In the following poem, "Minor Miracle," Nelson illustrates the feelings of vulnerability that she experiences, both as a cyclist and as a woman of color, during an encounter with locals in a small town in the Midwest. Do such encounters make people—especially women and people of color—less inclined to participate in activities that are physically vigorous and environmentally sustainable? Is the man's apology at the end of the poem sufficiently redemptive?

Which reminds me of another knock-on-wood
memory. I was cycling with a male friend,
through a small midwestern town. We came to a 4-way
stop and stopped, chatting. As we started again,
a rusty old pick-up truck, ignoring thc stop sign,
hurricaned past scant inches from our front wheels.
My partner called, "Hey, that was a 4-way stop!"
The truck driver, stringy blond hair a long fringe
under his brand-name beer cap, looked back and yelled,

"You fucking niggers!"
And sped off.
My friend and I looked at each other and shook our heads.
We remounted our bikes and headed out of town.
We were pedaling through a clear blue afternoon
between two fields of almost-ripened wheat
bordered by cornflowers and Queen Anne's lace
when we heard an unmuffled motor, a honk-honking.
We stopped, closed ranks, made fists.
It was the same truck. It pulled over.
A tall, very much in shape young white guy slid out:
greasy jeans, homemade finger tattoos, probably
a Marine Corps boot-camp footlockerful
of martial arts techniques.

"What did you say back there!" he shouted.
My friend said, "I said it was a 4-way stop.
You went through it."
"And what did I say?" the white guy asked.
"You said: 'You fucking niggers.' "
The afternoon froze.

"Well," said the white guy,
shoving his hands into his pockets
and pushing dirt around with the pointed toe of his boot,
"I just want to say I'm sorry."
He climbed back into his truck
and drove away.

John C. Ryan and Alan Thein Durning, "Bike (and Car)," from *Stuff: The Secret Lives of Everyday Things* (1997)

Alan Thein Durning, b. 1964; John C. Ryan

John C. Ryan and Alan Thein Durning's influential book *Stuff: The Secret Lives of Everyday Things,* published in 1997, follows a day in the life of a fictional, typical North American middle-class resident of Seattle. Tracing back the threads of distribution, commerce, and production involved in everyday consumer goods, *Stuff* examines the people and places affected every time the average American drinks a cup of coffee, clicks a mouse, or steps on the gas pedal. *Stuff* has even been used by mega-corporations such as Wal-Mart to educate their buyers about product life cycles.

Discussion Generators

In the excerpt that follows, the authors analyze a six-mile bicycle commute in an effort to understand its full implications: everything from the calories burned during the ride to the energy that went into the construction of the bike itself. What wider effects do your everyday actions and purchases have on other people and the natural environment? Does such a deep accounting practice affect your own decisions with regard to energy consumption? What other everyday objects, or "machines," might be particularly interesting to analyze in the context of energy?

This morning was warm and clear; I decided to bike to work. I strapped on my helmet and rolled off past the car left sitting in the driveway. I've done this before a few times and always felt better for it. But I keep falling back into driving. Maybe it's because my company gives me a free parking space for my car but no place to shower and change after biking.

Energy

I live six miles from my office, a good 20-minute workout on my bike. During the ride, I burned 210 calories—about what's in a plate of spaghetti, and less energy than any other form of transport consumes, even walking. If I had walked to work, I would have used 600 calories of energy—and gotten to work late.

If I had driven, I would have burned a quarter gallon of gasoline (from oil drilled on Alaska's North Slope, piped to a tanker in Prince William Sound, then shipped south and refined in Anacortes, Washington). That much gasoline would yield nearly 8,000 calories of energy—fossil fuel energy, not food energy—nearly 40 times the energy burned riding my bike. The main difference is that when I drive alone, 95 percent of the energy goes to moving the 3,200-pound car itself, not its 140-pound cargo. What a waste! On a bike, which weighs about one-fifth as much as I do, almost all the energy supplied by my muscles propels my body forward.

During the year, I will drive my car 11,600 miles—the average for an American—and buy 464 gallons of gasoline. I will also spend one-fourth of my personal income on transportation: 17 percent using my car, 7 percent on freight charges included in prices of things I buy, and 1 percent in taxes for roads. (My daily commute is shorter than most, but I drive a lot on weekend errands. Why do stores have to be scattered all over town? I also like to drive to my favorite mountain biking spots.)

Pollution

On my bike I caused no air pollution (unless you consider sweat air pollution) and made no contribution to global warming. I consumed no oil, gasoline, or other fossil fuels, and sent no toxic chemicals into the air. When I drive to work, my car pumps out nearly five pounds of climate-threatening carbon dioxide, half a pound of health-threatening carbon monoxide, and a few grams each of smog-forming hydrocarbons and nitrogen oxides.

Fossil fuels such as gasoline are the main source of climate-altering CO_2 emissions. In the United States, motor vehicles outnumber registered drivers, and Americans drive as many miles each year as all drivers in the rest of the world combined. Traffic accidents kill more Americans each year than guns or illegal drugs, and they are the leading cause of death for Americans 2 to 24 years old. Car exhaust also kills thousands of Americans annually.

Though my car has a catalytic converter to turn pollutants into CO_2 and water, it doesn't work well on such a short trip, when the engine runs cold. Half of all car trips in the United States are five miles or less—perfect for biking.

Pavement

The pavement I rode on was 12-inch-thick asphalt made from Texas petroleum poured over a graded roadbed covered with locally crushed rock. During Seattle's frequent rains, oil, road salt, and herbicides run off the pavement into Puget Sound. Nearly half of all cars on U.S. highways drip some hazardous fluid. About half the 20 quarts of oil my car used this year either was burned off by the car's engine or found its way onto the ground and into sewers and streams. (I can see the sheen of oil and antifreeze on my driveway.)

My bike took up a small fraction of the space my car takes on the roads and less than one-twentieth the area for parking. Bike lanes can move two to six times as many people per hour as the same area of pavement devoted to cars. Washington and Oregon already have more miles of roads than they do streams.

Steel

My bike weighs 30 pounds—mostly steel, aluminum, rubber, and plastics. The steel in my frame consists of iron with small amounts of carbon, chrome, and molybdenum added to make it harder. Such alloy steels are made in mini-mills that melt down scrap metal.

The 15 pounds of steel in my bike frame, wheels, and other parts began in a Chicago junkyard not far from the mini-mill. The scrap was sorted with magnets and delivered to the mill, where it was melted down in an electric-arc furnace. Three electrodes sent bolts of coal-fired electricity through the metal, and the intense heat melted the scrap. Impurities in my 15 pounds of scrap formed small amounts of gases, two ounces of toxic-laden dust, and a floating layer of waste called slag. Removing the waste generated a few grams of sludge tainted with heavy metals. Making steel from scrap uses one-fourth the energy of making steel from iron ore.

Iron Ore

Of the 1,800 pounds of steel in my car, 800 pounds began as scrap. Melting the scrap in an electric arc furnace generated 8 pounds of toxic dust. The other 1,000 pounds of steel and the 400 pounds of iron in my car came from a far dirtier process. It started with 3,500 pounds of iron ore in an open-pit mine—a giant red crater amid the spruce and pine forests of northern Minnesota. The ore was mined with gigantic Japanese-made machinery, crushed, ground, separated with magnets, and mixed with clay to form pellets. The 2,100 pounds of waste rock, or tailings, was dumped in a large pile near the mine. A barge carried the ore to Gary, Indiana, a steel town on the shores of Lake Michigan.

In a cavernous Gary steel mill, a blast furnace heated the ore with coke from West Virginia and limestone from Indiana to form pure molten iron. Impurities in the ore flowed off as iron-rich slag, which was later recycled into the blast furnace. Burning the coke also generated both carbon monoxide and carbon dioxide. The U.S. steel industry emits twice as much carbon monoxide as the pulp and paper industry, the next largest industrial source. (The largest source is not an industry; it's car exhaust.)

The molten iron was carried in specially lined railcars to the steelmaking furnace, where it was mixed with scrap steel from a salvage yard. Pure oxygen, blasted into the mix at supersonic speeds, kicked off a chemical reaction that melted the scrap and removed most impurities as slag. Forty-five minutes later, the iron and other inputs had been converted to steel—and to superheated clouds of zinc, lead, and iron oxide dust and more carbon monoxide. Water cooled this exhaust enough for smokestack scrubbers to remove some of the pollutants; these would be reused or trucked to landfills. Overall, the mill used 31 gallons—250 pounds—of water to make each pound of steel in my car.

All that steel makes my car heavy and strong. One reason I don't bike very often is that it's intimidating to face all those 3,000-pound steel boxes when my only protective gear is a one-pound helmet. But biking is actually safer per mile than driving, especially where there are good bike routes. Besides, if we all try to protect ourselves individually rather than collectively, we'll end up driving army tanks.

Paint

Once the steel for my bike had been molded into tubes, a truck took them to a bicycle factory in Wisconsin, where metal workers cut and welded them into a bike frame. Other workers sanded the frame, cleaned it with chemicals, oven-dried it, and sprayed on powdered paint (made of plastic resins and pigments in Ohio). The oversprayed powder fell through a grate in the floor and was collected for reuse. My bike frame was then baked to melt the paint into a uniform coat. Unlike liquid paint, powdered coatings contain no solvents, so they emit almost no air-polluting volatile organic compounds (VOCs) when sprayed; they also use less energy in baking. VOCs react with sunlight to form smog. Some liquid paint solvents also damage the kidneys and central nervous system and cause asthma, confusion, and fatigue.

My car's body was painted in a Detroit assembly plant. The body was dipped in detergent and a phosphate bath to remove surface debris, then in zinc phosphate and chromic acid to prevent corrosion and help the primer adhere to the metal. The body was submerged in a primer bath and baked, emitting VOCs.

Robots and workers then sprayed six more coats on my car: a sealant (consisting of polyvinyl chloride [PVC] and solvents), an anti-chip layer, a primer, a color coat, a clear coat, and a noise-reducing tar. The oversprayed liquids formed unusable sludges that were trucked to a landfill. Four times in seven coats, the body was baked, causing toxic VOCs to evaporate into the air. Painting is the most polluting process in the automobile assembly plant; the color coat emits the most air pollution.

Assembly

Once my bike frame was painted, it went to the assembly line where its various components were installed, mostly by hand. The frame, fork, and wheels were built in the Wisconsin plant; the rest were manufactured elsewhere, mostly in Japan.

The car assembly plant was a giant facility—as big as six football fields—housing a tremendously complex operation. At one end, robots started welding pieces of sheet metal; several hours and miles later, my completed car emerged. Welded in 4,000 spots, it had nearly 10,000 parts manufactured into about 100 major components by companies throughout the world. In all, making my car and its components consumed nearly 40,000 gallons of water, weighing nearly 100 times the car itself. My car's full life story could easily fill a book of its own.

Though making a car is energy intensive and polluting, most of my car's environmental impact comes when I sit behind the wheel. For example, making the car took roughly an eighth as much energy as the car will use during its nine-year lifetime. So my daily decisions about where to go and how to get there matter a lot. And my decisions about where to live matter even more. I know I've been driving less since I moved from the suburbs into Seattle.

Aluminum

My bike's eight pounds of aluminum were cast into gears, brakes, spokes, and other parts by a Japanese manufacturer. Smelters in Siberia made the metal from 40 pounds of Australian bauxite ore and hydroelectricity from a dam on Russia's Angara River, the outlet for Lake Baikal. The river was dammed to power aluminum smelters. The surface of the lake—the world's largest—rose three feet, flooding wetlands and island habitats of the nerpa, the world's only freshwater seal. The inefficient smelters have also severely polluted the Baikal region's air and water.

My car had many parts—pistons, wheels, the transmission—made at least partially of aluminum, by dozens of subcontractors in Asia and America. The 125 pounds of unrecycled aluminum began as 625 pounds of bauxite mined in Australia, Guinea, and Jamaica and smelted with hydropower from dammed rivers in Washington, Tennessee, Quebec, and Russia. One car wheel alone contained more aluminum than the entire bicycle.

Synthetics

My bicycle had nylon cable guides made in Delaware; polyurethane handlebar grips made in Italy; a vinyl and polyurethane seat; and knobby, mud-splattered tires of butadiene rubber from Taiwan—in all, about eight pounds of synthetic materials.

After the car body was painted, unionized autoworkers earning $16 an hour installed "hard trim" such as windshields, windows, and mirrors (90 pounds of glass made in Pittsburgh and Nashville). Then they glued and snapped in "soft trim." This included the instrument panel (PVC made in Delaware and electronics from Japan), door pads (polypropylene from New Jersey), seats (steel, foam, and vinyl from Illinois), and carpets (sewn in Georgia of nylon and rayon from various chemical plants). Each year, U.S. auto manufacturers consume 683 million pounds of petrochemical glues and adhesives plus 135 square miles of carpet, enough to cover the city of Detroit.

Rubber

Probably the biggest use of synthetic chemicals in my car was in the tires. They were manufactured in Alabama of steel wires; rayon fabric synthesized from Arkansas pine pulp; and rubber made from Mexico petroleum, Louisiana sulfur, and various additives.

My car contained 136 pounds of synthetic rubber, the bicycle about 5. Some 70 percent of synthetic rubber produced worldwide becomes car tires and other auto parts. Landfills, vacant lots, and ravines in the United States hold about 2 billion old car tires. That's eight for me and eight for every other person in the country. More than 200 million tires are added to the pile each year.

My bike and my car rolled off their respective assembly lines and were trucked to the Seattle area. Nine cars fit on the truck from Detroit; 500 bikes in cardboard boxes fit on the truck from Wisconsin. The boxes would later be recycled, thrown out, or reused by University of Washington students shipping their bikes home at the end of the school year.

William Stafford, "Maybe Alone on My Bike" (1964)

1914–1993

William Stafford was born in Hutchinson, Kansas, but spent much of his life in Lake Oswego, Oregon, teaching at Lewis & Clark College from 1948 until his retirement. He published numerous volumes of poetry, receiving the National Book Award in 1963 for *Travelling Through the Dark*. His nonfiction memoir, *Down in My Heart*, recounts his experiences as a conscientious objector during World War II. "Maybe Alone on My Bike," which first appeared in *The New Yorker* in April 1964 and was later collected in Stafford's 1983 book *Smoke's Way*, celebrates one of humanity's most efficient and joyous technologies—the bicycle.

Discussion Generators

Energy-conscious readers can imagine the speaker of the poem "quaintly on a cold / evening pedaling home" from work, experiencing the world in a way that's out of reach to throngs of commuters in automobiles. What does this poem have to say about the relationship between people and machines (and the appropriate use of our muscle energy)? How might this relationship influence our appreciation of the larger world?

I listen, and the mountain lakes
hear snowflakes come on those winter wings
only the owls are awake to see,
their radar gaze and furred ears
alert. In that stillness a meaning shakes;

And I have thought (maybe alone
on my bike, quaintly on a cold
evening pedaling home), Think!—
the splendor of our life, its current unknown
as those mountains, the scene no one sees.

O citizens of our great amnesty:
we might have died. We live. Marvels
coast by, great veers and swoops of air
so bright the lamps waver in tears,
and I hear in the chain a chuckle I like to hear.

Rhiannon McClatchey, "Kinesis" (2014)

b. 1987

Freelance blogger and poet Rhiannon McClatchey draws upon the spectacular surroundings of her home in Colorado's Front Range to reflect upon her experiences as a long-distance trail runner. Her poetry explores relationships between feminism, athletics, and environmentalism, imagining what might be possible were Western culture to understand people—particularly women—as active subjects rather than passive objects. McClatchey's word play and metaphors invite readers to discover relationships between the human body and the earth, and in doing so she considers how such language might heal both.

Discussion Generators

Many of the discussions and debates about energy center upon various energy regimes: fossil fuels, nuclear power, hydroelectricity, solar, and the like. An often overlooked source of energy, however, is the human body itself. In what ways might the power contained within our own bodies reduce our dependence upon these other sources of energy? Might greater awareness of our own physical potential enable us to rely less on technological solutions? McClatchey's poem deploys words with multiple meanings to help readers make connections that might not seem immediately obvious. Which word choices do you notice, and how do they encourage readers to think differently about their own bodies in relation to the natural environment?

> Mountain paths have memorized my step. My running cadence is a key the soundscape kept, and as I am fleet, I harmonize my breath with place, in time. My muscles shape the beat. I feel my life has purpose and is safe in song.
>
> When I inhabit my body, I sing my native language. I remember who I am, where I should come from and what I'm meant to do. My mother first saw me in a sonogram. She knew me by my vibration. My being was a kick. When I entered earth, my first words were movement. When I landed, I was kinetic. By now, I intuit a body hides reserves you can't use looking outside in. I have to start inside, as subject, to release the memory of my power.
>
> Endurance is the ability to locate energy the world would have me waste. Running on trails, I am fluent in myself and fluid in my body. I drink from wellsprings of will.
>
> I don't burn calories to starve into lovely static. I run from a world that shames my excess, asks me to languish, to stay put, so they can get a good look. I run in an environment that encourages my first purpose, which is to be dynamic, *to do something* with what I have. And to nourish myself, so I can do it again and again.

Sheryl St. Germain, "Midnight Oil" (2011)

b. 1954

New Orleans and its rich Mississippi Delta hold lasting importance to Louisiana native Sheryl St. Germain, who has only intensified her connections to this place after the devastation of Hurricane Katrina and the BP oil disaster. St. Germain is the author of several books of poetry and nonfiction essays, including *The Mask of Medusa* (1987), *Making Bread at Midnight* (1992), *How Heavy the Breath of God* (1994), *Swamp Songs: The Making of an Unruly Woman* (2003), and *Let It Be a Dark Roux: New and Selected Poems* (2007). She has been the recipient of two NEA Fellowships, an NEH Fellowship, the Dobie-Paisano Fellowship, and the William Faulkner Award for the personal essay. St. Germain directs the Master of Fine Arts program in creative writing at Chatham University in Pittsburgh, which concentrates on nature, environment, and travel writing.

Discussion Generators

In her poem "Midnight Oil," St. Germain faces her anguish at the 2010 Deepwater Horizon oil disaster in the Gulf. In her reaction to the polluted coastline, she attempts to accept her share of blame, writing, "Let's ask those responsible / and some of those are us // to walk deep out into the waters / . . . let them try to explain." While the poet suggests citizens such as herself are partly "responsible" for the BP oil spill, the poem is intentionally slippery. Consider all the different movements and reactions that the oil and the Gulf power—human, corporate, and animal. How does the motion of the poem ebb and flow in response to the many actors with a stake in this volatile mix of water and oil?

how to speak of it
this thing that doesn't rhyme
or pulse in iambs or move in predictable ways
like lines
or sentence

how to find the syntax
of this thing
that rides the tides
and moves with the tides and under the tides
and through the tides
and has an underbelly so deep and wide
even our most powerful lights
cannot illuminate its full body

this is our soul shadow,
that darkness we cannot own
the form we cannot name

and I can only write about it at night
when my own shadow wakes me, when I can feel
night covering every pore and hair follicle, entering eyes
and ears, entering me like Zeus, a night I don't want
on me or in me, and I dream of giving birth
to a rusty blob of a child who slithers out of me,
out and out and won't stop slithering, growing and darkening,
spreading and pulsing between my legs
darkening into the world

---

what it might feel like to be a turtle, say,
swimming in the only waters you have ever known
swimming because it is the only way you move through the world
to come upon this black bile
a kind of cloying lover

a thing that looks to you
like a jellyfish, so you dive into it and try to eat it
but it covers your fins so they can't move as before
and there is a heaviness on your carapace and head
that wasn't there before, and you are blind
in the waters of your birth

When the summers got too hot even for those of us born in New Orleans, so hot that our ancestors' bones sweated and complained in their vaults, my father would decide it was time, and the family would pile into the station wagon and drive down to Grand Isle, where we'd run along the beach into the Gulf as if into a lover's arms, smash into the salty waves, swim until the sun went down and we were red as boiled crawfish.

Mother would have made a pungent crab salad, with quarters of crab marinated in garlic and olive oil, lemon and celery. Sometimes we'd have boiled shrimp or crabs. The grownups would stay up drinking and playing cards at night while we slept the sweet sleep of children who don't yet know what stygian rivers run in their veins. Exhausted from swimming, hair still damp and smelling like the Gulf, we'd huddle together in the big bed on the screened porch. The smell and sound of the waves rocked us to sleep, dreams of pelicans and gulls and flying fish filled our heads and hearts, and we were content.

On rainy days when we couldn't swim my father taught me how to play pool in one of the hulking bars that used to front the beach. The bar's gone now, like the house we stayed in, destroyed by hurricanes that wipe out

every human-made thing every few years.

Oiled Birds, ***edited from Wikipedia:***
Penetrates plumage, reduces insulating ability, makes birds vulnerable to temperature fluctuations, less buoyant in water. Impairs bird's abilities to forage and escape predators. When preening, bird ingests oil that covers feathers, causing kidney damage, altered liver function, digestive tract irritation. Foraging ability is limited. Dehydration, metabolic imbalances. Bird will probably die unless there is human intervention.

I'm looking at an old photo of my brother, right after he got out
of prison. He's twenty, sitting on a beach chair at Grand Isle,
looking gaunt and pale, but smiling.
It was the first place my mother thought to take him
when she feared the grime and shame of that other place
might have tarnished his heart too deeply.

She knew he loved this island, where simple things
like saltwater and clean beaches,
birds and fish, crabs,
might act like containment booms,
keep the demons away.

He'd die a few years later,
his liver polluted with what he thought
would make the world bearable,
and a few years after him
my father would go
from that same staining.

Now, when I look at these beaches I love,
greasy with oil as far as I can see,
when I think of how this island
and its marshes should act like filters,
I think of my father and brother,
I think this is what
their livers must have looked like
as they moved toward the end, darkening,
becoming pebbly with disease, finally
too black with blight
to filter
anything.

It's June in Pittsburgh where I live now, hot and muggy,
and it feels like a day my father would've said *let's go*.

People don't want to look at the pictures anymore of the birds
and turtles, the fish, the oiled beaches
they want to go on to something else
they don't want to hear about the old fishermen
who may never fish again
the ones being trained to clean up instead of fish

It's an old story, really, how we always dirty what we love,
and I'm tired too, have seen way too many pictures
of oiled birds and the oiled waters of this dear place

and I've heard way too many pundits and politicians
and newsmen analyze, blame and predict

and jokemen joke:
let's call the Gulf *the Black Sea*.

Dear CNN: even the devil would bore us
if he was on 24 hours a day

there are times we need silence
as much as we need news

or a poem that creates a silence
in us where we can feel again

How people from Louisiana have described the oil:
Brown and vivid orange globs. Tar balls. Thick gobs. Red waves. Deep stagnant ooze. Clumps of tar. Consistency of latex paint. Sheets of foul-smelling oil. Patches of oil. Caramel-colored oil. Tide of oil. Red brown oil. Rainbows of Death. Waves of gooey tar blobs.

Bruised internal organs of a human body. Heavy heavy slickoil. Oil sheen. Oily stench. Melted chocolate.

An eye for an eye, my father might say,
a tooth for a tooth.

Let's ask those responsible,
and some of those are us

to walk deep out into the waters
of this once beautiful island,
the waters that once teemed with speckled trout,
oysters, shrimp,
let's ask them to walk far out into it,
to swim out with long sweeping strokes,

and then,
when they are thick and covered
with the stuff, when it's in their hair and blinding
them, stopping up their ears and mouths,
when it's sticking to every pore
in their body,

then
let them try to swim back

then
let them try to explain

Sinclair Lewis, from *Main Street* (1920)

1885–1951

The works of Nobel Prize–winning novelist and playwright Sinclair Lewis are known for their insightful and critical views of American society and capitalism, as well as for their strong characterizations of modern workers. Born in the village of Sauk Centre, Minnesota, Lewis set several of his novels, including *Main Street* (1920)—his first novel and a significant commercial success—in his native state. When the book was published, it was common for Americans to valorize purportedly wholesome small towns like the fictional Gopher Prairie, a notion satirized by the vicious realism and biting humor in *Main Street.*

Discussion Generators

Americans once depended upon trains for goods from faraway places, and they imagined the trains as a means of escape from their humdrum lives. What are some ways that Lewis articulates the symbolic meaning that trains occupy in the imaginations of Gopher Prairie's residents? What might life in America be like if trains, rather than highways, were still the primary arteries of transportation?

Trains!

At the lake cottage she missed the passing of the trains. She realized that in town she had depended upon them for assurance that there remained a world beyond.

The railroad was more than a means of transportation to Gopher Prairie. It was a new god; a monster of steel limbs, oak ribs, flesh of gravel, and a stupendous hunger for freight; a deity created by man that he might keep himself respectful to Property, as elsewhere he had elevated and served as tribal gods the mines, cotton-mills, motor-factories, colleges, army.

The East remembered generations when there had been no railroad, and had no awe of it; but here the railroads had been before time was. The towns had been staked out on barren prairie as convenient points for future train-halts; and back in 1860 and 1870 there had been much profit, much opportunity to found aristocratic families, in the possession of advance knowledge as to where the towns would arise.

If a town was in disfavor, the railroad could ignore it, cut it off from commerce, slay it. To Gopher Prairie the tracks were eternal verities, and boards of railroad directors an omnipotence. The smallest boy or the most secluded grandma could tell you whether No. 32 had a hot-box last Tuesday, whether No. 7 was going to put on an extra daycoach; and the name of the president of the road was familiar to every breakfast table.

Even in this new era of motors the citizens went down to the station to see the trains go through. It was their romance; their only mystery besides mass at the Catholic Church; and from the trains came lords of the outer world—traveling salesmen with piping on their waistcoats, and visiting cousins from Milwaukee.

Gopher Prairie had once been a "division-point." The roundhouse and repair-shops were gone, but two conductors still retained residence, and they were persons of distinction, men who traveled and talked to strangers, who wore uniforms with brass buttons, and knew all about these crooked games of con-men. They were a special caste, neither above nor below the Haydocks, but apart, artists and adventurers.

The night telegraph-operator at the railroad station was the most melodramatic figure in town: awake at three in the morning, alone in a room hectic with clatter of the telegraph key. All night he "talked" to operators twenty, fifty, a hundred miles away. It was always to be expected that he would be held up by robbers. He never was, but round him was a suggestion of masked faces at the window, revolvers, cords binding him to a chair, his struggle to crawl to the key before he fainted.

During blizzards everything about the railroad was melodramatic. There were days when the town was completely shut off, when they had no mail, no express, no fresh meat, no newspapers. At last the rotary snow-plow came through, bucking the drifts, sending up a geyser, and the way to the Outside was open again. The brakemen, in mufflers and fur caps, running along the tops of ice-coated freight-cars; the engineers scratching frost from the cab windows and looking out, inscrutable, self-contained, pilots of the prairie sea—they were heroism, they were to Carol the daring of the quest in a world of groceries and sermons.

To the small boys the railroad was a familiar playground. They climbed the iron ladders on the sides of the box-cars; built fires behind piles of old ties; waved to favorite brakemen. But to Carol it was magic.

She was motoring with Kennicott, the car lumping through darkness, the lights showing mud-puddles and ragged weeds by the road. A train coming! A rapid chuck-a-chuck, chuck-a-chuck, chuck-a-chuck. It was hurling past—the Pacific Flyer, an arrow of golden flame. Light from the fire-box splashed the under side of the trailing smoke. Instantly the vision was gone; Carol was back in the long darkness; and Kennicott was giving his version of that fire and wonder: "No. 19. Must be 'bout ten minutes late."

In town, she listened from bed to the express whistling in the cut a mile north.

Uuuuuuu!—faint, nervous, distrait, horn of the free night riders journeying to the tall towns where were laughter and banners and the sound of bells—Uuuuu! Uuuuu!—the world going by—Uuuuuuu!—fainter, more wistful, gone.

Down here there were no trains. The stillness was very great. The prairie encircled the lake, lay round her, raw, dusty, thick. Only the train could cut it. Some day she would take a train; and that would be a great taking.

Bill McKibben, "A Moral Atmosphere: Hypocrisy Redefined for the Age of Global Warming" (2013)

b. 1960

"Our comforting sense of the permanence of our natural world, our confidence that it will change gradually and imperceptibly if at all, is the result of a subtly warped perspective," writes Bill McKibben in his provocative 1989 book *The End of Nature,* which argues that human-produced phenomena such as the greenhouse effect, acid rain, and the depletion of the ozone layer have caused such radical change in the earth's atmosphere and climate system that the very idea of nature as we know it has ceased to exist. McKibben has spent most of his professional career writing about such topics as global climate change, alternative energy, and the ethics of genetic engineering. He is the founder of the group 350.org, which seeks to build a global grassroots movement to solve the climate crisis. The name derives from the group's effort to reduce the amount of carbon dioxide in the atmosphere to below 350 parts per million.

Discussion Generators

In the essay that follows, McKibben—who has traveled widely to give talks and presentations on climate change—explains the rationale given by some Americans for their opposition to government, or even individual, action on climate change. This argument, according to McKibben, amounts to accusing climate change activists of hypocrisy. How valid does McKibben believe this sort of criticism is? To what extent do you think it is appropriate to hold activists to higher standards of environmentally sustainable behavior than the general population? If activists were to find ways to leave a smaller carbon footprint, would this increase the force of their arguments? Does McKibben have a point when he says that the increased scrutiny on individual behaviors actually points out that the problem is with the overall system, not with individuals?

The list of reasons for not acting on climate change is long and ever-shifting. First it was "there's no problem"; then it was "the problem's so large there's no hope." There's "China burns stuff too," and "it would hurt the economy," and, of course, "it would hurt the economy." The excuses are getting tired, though. Post Sandy (which hurt the economy to the tune of $100 billion) and the drought ($150 billion), 74 percent of Americans have decided they're very concerned about climate change and want something to happen.

But still, there's one reason that never goes away, one evergreen excuse not to act: "you're a hypocrite." I've heard it ten thousand times myself—how can you complain about climate change and drive a car/have a house/turn on a light/raise a child? This past fall, as I headed across the country on a bus tour to push for divestment from fossil fuels, local newspapers covered each stop. I could predict, with great confidence, what the first online comment from a reader following each account would be: "Do these morons not know that their bus takes gasoline?" In fact, our bus took biodiesel—as we headed down the East Coast, one job was watching the web app that showed the nearest station pumping the good stuff. But it didn't matter, because the next comment would be: "Don't these morons know that the plastic fittings on their bus, and the tires, and the seats are all made from fossil fuels?"

Actually, I do know—even a moron like me. I'm fully aware that we're embedded in the world that fossil fuel has made, that from the moment I wake up, almost every action I take somehow burns coal and gas and oil. I've done my best, at my house, to curtail it: we've got solar electricity, and solar hot water, and my new car runs on electricity—I can plug it into the roof and thus into the sun. But I try not to confuse myself into thinking that's helping all that much: it took energy to make the car, and to make everything else that streams into my life. I'm still using far more than any responsible share of the world's vital stuff.

And, in a sense, that's the point. If those of us who are trying really hard are still fully enmeshed in the fossil fuel system, it makes it even clearer that what needs to change are not individuals but precisely that system. We simply can't move fast enough, one by one, to make any real difference in how the atmosphere comes out. Here's the math, obviously imprecise: maybe 10 percent of the population cares enough to make strenuous efforts to change—maybe 15 percent. If they all do all they can, in their homes and offices and so forth, then, well . . . nothing much shifts. The trajectory of our climate horror stays about the same.

But if 10 percent of people, once they've changed the light bulbs, work all-out to change the system? That's enough. That's more than enough. It would be enough to match the power of the fossil fuel industry, enough to convince our legislators to put a price on carbon. At which point none of us would be required to be saints. We could all be morons, as long as we paid attention to, say, the price of gas and the balance in our checking accounts. Which even dummies like me can manage.

I think more and more people are coming to realize this essential truth. Ten years ago, half the people calling out hypocrites like me were doing it from the left, demanding that we do better. I hear much less of that now, mostly, I think, because everyone who's pursued those changes in good faith has come to realize both their importance and their limitations. Now I hear it mostly from people who have no intention of changing but are starting to feel some psychic tension. They feel a little guilty, and so they dump their guilt on Al Gore because he has two houses. Or they find even lamer targets.

For instance, as college presidents begin to feel the heat about divestment, I've heard from several who say, privately, "I'd be more inclined to listen to kids if they didn't show up at college with cars." Which in one sense is fair enough. But in another sense it's avoidance at its most extreme. Young people are asking college presidents to stand up to oil companies. (And the ones doing the loudest asking are often the most painfully idealistic, not to mention the hardest on themselves.) If as a college president you *do* stand up to oil companies, then you stand some chance of changing the outcome of the debate, of weakening the industry that has poured billions into climate denial and lobbying against science. The action you're demanding of your students—less driving—can't rationally be expected to change the outcome. The action they're demanding of you has at least some chance. That makes you immoral, not them.

Yes, they should definitely take the train to school instead of drive. But unless you're the president of Hogwarts, there's a pretty good chance there's no train that goes there. Your students, in other words, by advocating divestment, have gotten way closer to the heart of the problem than you have. They've taken the lessons they've learned in physics class and political science and sociology and economics and put them to good use. And you—because it would be uncomfortable to act, because you don't want to get crosswise with the board of trustees—have summoned a basically bogus response. If you're a college president making the argument that you won't act until your students stop driving cars, then clearly you've failed morally, but you've also failed intellectually. Even if you just built an energy-efficient fine arts center, and installed a bike path, and dedicated an acre of land to a college garden, you've failed. Even if you drive a Prius, you've failed.

Maybe especially if you drive a Prius. Because there's a certain sense in which Prius-driving can become an out, an excuse for inaction, the twenty-first-century equivalent of "I have a lot of black friends." It's nice to walk/drive the talk; it's much smarter than driving a semi-military vehicle to get your groceries. But it's become utterly clear that doing the right thing in your personal life, or even on your campus, isn't going to get the job done in time; and it may be providing you with sufficient psychic comfort that you don't feel the need to do the hard things it will take to get the job done. It's in our role as citizens—of campuses, of nations, of the planet—that we're going to have to solve this problem. We each have our jobs, and none of them is easy.

Don McKay, "Nocturne Macdonald-Cartier Freeway" (2004)

b. 1941

Bird themes and flight are dominant topics in the poetry of Canadian poet and avid birdwatcher Don McKay. In *Birding, or Desire* (1983), the quirky protagonist is never far from his *Birds of Canada* field guide. McKay sees his writing as "nature poetry in a time of environmental crisis," and his message is always ecological, inspired by the conflict between inspiration and spirituality, instinct and knowledge. Other key members of this Canadian group of "ecopoets" include Tim Lilburn, Dennis Lee, Roo Borson, Robert Bringhurst, and Jan Zwicky.

Discussion Generators

In "Nocturne Macdonald-Cartier Freeway," McKay attempts to reconcile the widespread perception of human isolation from the natural world with his firm belief that we never truly exist apart from the laws of ecology, evolution, physics—and, yes, birds. If humans are inexorably a part of the natural world, why then does it require so much mental effort to remind ourselves of this basic fact? What does all of this have to do with driving, with the machinery of automobiles? What do you think McKay means when he writes, "We drive because we believe in the death of traffic"?

In the archipelago of coffee, each man is an island. The women are—who knows where—withdrawn but not quite vanished, like god at the end of the nineteenth century. Between your figure and the ground there is a tissue of airless space about the thickness of a piece of paper, in which all double helices untwine, adieu my little corkscrew, and swim offstage at the speed of light. Warnings, some visible, are posted at each junction. The floor may be slippery, the eyes in the mirror may be holes, the cashier may be unfamiliar with your gravity, the money may be avian. But the coffee is real and powers the economy.

Along the re-entry ramp the transports twinkle. Probably their drivers are asleep, ghosts in the machines. We say goodbye to Christmas in Cubism and follow our headlights into the dark.

We drive because we believe in the death of traffic. There will be a kitchen in the middle of a forest, its windows widening slowly, reaching their frames but continuing until the walls are erased. You turn on the tap, an underground river leaps sixty feet into your mouth, a perfectly composed dream. No phone-ins. No hits from the sixties. No eye in the sky. No internal combustion of any kind. No memory lane. The first song sparrow will have your whole head to itself.

David Mas Masumoto, "Callous Hands," from *Harvest Son: Planting Roots in American Soil* (1998)

b. 1954

David Mas Masumoto is a third-generation Japanese American. He calls himself an "artist farmer" who engages the world through "literary farming." The owner and operator of a certified organic farm and the author of several books about his experiences as a farmer and Japanese American, including *Epitaph for a Peach* and *Four Seasons in Five Senses*, Masumoto lives with his family twenty miles south of Fresno, California.

Discussion Generators

Energy on the farm, Masumoto argues in the essay below, has traditionally been exerted through the bodies of farmers, creating "fortune" through calluses and muscles. What are the implications for energy use in the cultural transition from Masumoto's grandparents' generation to his own? Modern technology may save wear and tear on farmers' bodies, but does it result in more efficient application of energy resources?

I once met an older Nisei and he asked where my grandparents came from. I answered, "Kumamoto and Hiroshima." He grinned and shook my hand again, only harder. "Hiroshima good, but *Kumamoto* the best. We country people know how to work!" As his grip tightened, I could feel his dry, rough palms scraping my skin.

I imagine the hands of my grandparents to be quite hard, their thick calluses badges of honor earned only after years in the fields, a carefully engraved character in the flesh of *inaka*—a term Japanese "country folk" call themselves, quite conscious of the pejorative connotation of "peasant." The hands tell a story of worth, private statements like a *haiku* I would have written were I an immigrant—

Thick hands work
Under a harvest sun
Fortune with another callus!

But the calluses of Issei would have been hidden. The men bow and infrequently shake hands; the women smile and drop their heads, acts of humility that mask the coarseness folded in their laps. Their hands embrace a silence, their personal struggle to create a home in a new land.

I'm sure my grandparents also wore farmers' tans, dark arms and faces with a dividing line along the neck and the arms where collar and shirtsleeves stop. Both my grandmothers tried to hide their faces and arms from the sun by wearing bonnets and fore-arm coverings made from used rice sacks. The coarse cotton bags provided excellent protection. A Nisei farm wife described how a bonnet, with its bleached and faded white material, helped reflect the heat and the long sides and back panel blocked out sunlight. "For a brim, Chinese rice sacks worked better"—the material was stiff and made a better edge. Generations of Japanese farm women wore such bonnets, guarding against the intense rays, trying to maintain a delicate and pale face. Some succeeded, but most, like Baachan Masumoto, eventually gave up. The best and thickest bonnets could not prevent a dark tan and weathered complexion after hours and hours laboring in the sun.

When I first returned to the farm, I did not tan well. At first I worked without a shirt and my back stung from the exposure. I quickly learned to protect myself and wore T-shirts and work-shirts. Then on my hands the blisters began to appear. I tried to ignore the tingling patches of red skin until the bulbous pockets of flesh and fluid had swollen to near bursting. I jabbed at them, surprised by their sudden appearance, entranced as I poked them and felt the tickle of pain with my skin stretched and pulsating. I had no choice but to return to work. The blisters were quickly popped and shredded; the clear liquid trickled over the tingling surface, the tenderness stinging for days before new skin grew and I could peel away the dead tissue. Even after a week, I could only delicately pick up a shovel or gingerly grip the tractor steering wheel. My raw hands lacked time in the orchards and vineyards.

Gradually I grew tougher, yet I never established the same thick calluses as my dad or Baachan. Though I prune trees, tie vines, and dig out weeds, the physical work of my generation differs from the past. I have replaced a strong back with modern equipment and technology. Many times I find myself acting more like businessman than farmer, seeking bottom-line profits, accumulating capital, longing to expand my operation. Yet my hunger for success is not satisfied by money alone. I long to learn both how to farm and how to enjoy being a son, brother, husband, and father. I do not yet have answers and want to continue my journey. I realize my hands do not have the required history; only time and memory will bring me the success I am after.

I never touched the hands of my grandfathers. I only know the ragged fingertips and uneven palms of my grandmothers, especially Baachan Masumoto, who sometimes would *momo*/massage our backs.

She walks with old bony legs that shuffle through garden weeds. Her four-foot-eight-inch frame slouches forward as if walking against the wind. She still lugs buckets of fruit and vegetables from her garden, still possesses a strength earned by years of farmwork.

When we were children, she teased us kids by showing us her firm biceps. We asked if there was an egg underneath her skin. Mom translated and Baachan laughed out loud, one of the few times she did so. Then she clenched her fist tighter and the "egg" contracted, becoming rock-hard. We screamed and roared, then whispered to each other, "You touch it. You touch it." Eventually we bravely reached out and tapped her muscle, our eyes wide with amazement as we agreed that there was an egg in there. A chorus of more laughs and screams followed.

Baachan has worked in the fields for her entire life. The long years are manifested in her hands, gnarled with calluses, coarse with dry cracks in the hardened tips of her thumbs and index fingers. Her fingernails have yellowed, the flesh underneath long ago dried out, and appear brittle. Yet these are the same hands that become filled with a warmth as she massages our backs.

We kids called it "to *momo*," and one of the few Japanese phrases we knew was "Baachan, *momo*, *momo* us please!" We stood with our backs to her, our young, tanned bodies next to our gray-haired, wrinkled grandmother. She began slowly, moving in a circular motion on each shoulder blade, then randomly dancing across our skin, gently pressing with her hands. The calluses slid over our bones and muscles and generated heat. I closed my eyes, feeling her hand across my back, giggling when my sister or brother tried to push me out of the way to get their turn with Baachan. When her arm grew tired, she switched hands and renewed the warmth. I sometimes sighed out loud and she changed from rubbing to softly tapping. I would take a deeper breath and renew my sigh, listening to the pulsating sound created by air nudged from my lungs. We created a duet, a chant, a Japanese-American song of generations.

As we grew older, Baachan massaged the kids less and less, but every once in a while I feel her old farm hands. Since I have returned, I often think about her time as a worker. I have benefited much from each of the scars in her hands. As she begins to rub my skin, I can feel the coarseness of her memories. She has contributed to who and what I can become. I celebrate her life now that I too work the land. As her old hands playfully rub my young back, a soothing history passes between generations.

For my grandparents to succeed in America, they were forced to have callous hands. They believed that hard work would be rewarded and accepted their place, toiling on someone else's land, saving until they could buy their own.

A. R. Ammons, "Mechanism" (1960)

1926–2001

A. R. Ammons grew up on a tobacco farm near Whiteville, North Carolina. His experiences there during the Great Depression inspired a great deal of his poetry, which he began writing while serving aboard a Navy destroyer during World War II. After the war, he completed his education at Wake

Forest University and the University of California at Berkeley, and he then worked as an elementary school principal and glass salesman before beginning his teaching career at Cornell University in 1964. He won the National Book Award twice: in 1973 for *Collected Poems 1951–1971*, and again in 1993 for *Garbage*. In poems such as "Cascadilla Falls," "The Wide Land," "Poetics," and many others, Ammons articulates the tension between the individual's sense of self as bound within space and time, and the sense of self as part of a larger order. On Ammons's composition process, literary critic Harold Bloom has observed that Ammons "compar[es] the energy of his poems to the energy of nature," a relationship that can be seen vividly in "Mechanism," the poem reprinted below.

Discussion Generators

In the first stanza of "Mechanism," Ammons lists four items—"goldfinch, corporation, tree, / morality"—as examples of what he calls a "going thing." What characteristics could these four things possibly have in common? Goldfinches and trees are both living organisms, of course, but what about corporations and morality? How do the "incoming and outgoing energies" of these four nouns compare to energy regimes involving machines and fossil fuels? The second through fourth stanzas use considerable repetition of the words "energy" or "energies" to set forth the meaning and tone of the poem. What kinds of "energies" is Ammons referring to here? In what ways might Ammons's poem broaden our sense of what "energy" is?

Honor a going thing, goldfinch, corporation, tree,
morality: any working order,
animate or inanimate: it

has managed directed balance,
the incoming and outgoing energies are working right,
some energy left to the mechanism,

some ash, enough energy held
to maintain the order in repair,
assure further consumption of entropy,

expending energy to strengthen order:
        honor the persisting reactor,
    the container of change, the moderator: the yellow

bird flashes black wing-bars
        in the new-leaving wild cherry bushes by the bay,
    startles the hawk with beauty,

flitting to a branch where
        flash vanishes into stillness,
    hawk addled by the sudden loss of sight:

honor the chemistries, platelets, hemoglobin kinetics,
        the light-sensitive iris, the enzymic intricacies
    of control,

the gastric transformations, seed
        dissolved to acrid liquors, synthesized into
    chirp, vitreous humor, knowledge,

blood compulsion, instinct: honor the
        unique genes,
    molecules that reproduce themselves, divide into

sets, the nucleic grain transmitted
        in slow change through ages of rising and falling form,
    some cells set aside for the special work, mind

or perception rising into orders of courtship,
        territorial rights, mind rising
    from the physical chemistries

to guarantee that genes will be exchanged, male
        and female met, the satisfactions cloaking a deeper
    racial satisfaction:

heat kept by a feathered skin:
the living alembic, body heat maintained (bunsen
burner under the flask)

so the chemistries can proceed, reaction rates
interdependent, self-adjusting, with optimum
efficiency—the vessel firm, the flame

staying: isolated, contained reactions! the precise and
necessary worked out of random, reproducible,
the handiwork redeemed from chance, while the

goldfinch, unconscious of the billion operations
that stay its form, flashes, chirping (not a
great songster) in the bay cherry bushes wild of leaf.

SECTION TWO

energy debates

Introduction to Energy Debates

Mention the topic of energy, and most people will expect the conversation to leap immediately to the cost of monthly home heating bills, the price of gas at the local pumps, or debates about the cleanest, cheapest, and most "independent" energy resources of the future. This anthology begins with a more philosophical focus on the essential nature of energy, separate from the daily energy debates and controversies that occupy the news media. The next two chapters, however, turn directly to such familiar topics as fossil fuels, nuclear energy, and alternative (renewable) energy resources.

Chapter Three explores the *fossil* nature of so-called fossil fuels and the complications of living during the "Atomic Age." Many citizens know the phrase "fossil fuels" from the news but don't stop to consider the geological processes that create oil and coal or the implications of mining these energy sources from deep underground. But the debates concerning fossil fuels occur not only between "environmentalists" and those, at the other extreme, who derive their incomes directly from exploiting these resources. While Sandra Steingraber shudders at the prospect of disrupting the ecology of upstate New York by industrializing the rural landscape, Drum Hadley, who has long worked to conserve habitat in the Southwest, finds that machine operators have something to say, and we'd do well to pay attention to those cadences as well. The reality is that the apparent disagreement between these two writers plays out in the minds of each of us who uses fossil fuels (all of us) and cares about living on a healthy planet (this would probably also include all readers).

The mixed feelings about fossil fuels also occur in the context of nuclear power. While writers like Rebecca Solnit and Wendy Rose raise alarm about the dangers of radiation and the imperiousness of the nuclear industry, others such as Marilou Awiakta express wonderment at the mysterious power of the atom. What's clear is that nuclear energy is no panacea, no easy solution to the climate concerns prompted by the burning of fossil fuels. The misgivings of the atomic age are just as intense in the twenty-first century as they were at the dawn of this technology in the mid-twentieth century.

Even as we continue to burn fossil energy and split atoms for power, modern societies are becoming increasingly attuned to the virtues of cleaner, cheaper, and more durable alternatives, as the selections in Chapter Four directly or indirectly suggest—of course, these alternatives might also include conservation (or simply finding strategies to use fewer external energy sources to sustain our ways of life). As the excerpt from George Eliot's 1860 novel *Mill on the Floss* reveals, the use of water and other natural forces as a source of energy is not simply an outgrowth of current concerns about acid rain, global warming, and nuclear radiation. Many of the texts in this chapter do not explicitly probe the pros and cons of using wind, sunlight, or geothermal energy as practical solutions to today's energy needs. Instead, being works of literature, they contemplate the idea and experience of these phenomena through images and stories, spurring readers to debate what it might mean to employ such forces in our everyday lives.

The earth's body and our own bodies are susceptible to the consequences of careless and excessive energy use. The literary texts in Chapter Five shed light on the often invisible costs of our energy-consuming lives. The reality is that our small, individual lives have a forceful collective impact on the planet and on the lives of other beings, unseen and far away. We participate in the depletion of energy resources and in the degradation of the natural world through our rampant hunger for power.

While many of the selections throughout this anthology directly confront energy-related topics familiar through the popular media, other pieces open up the mysterious and beautiful connections between our lives and the planet. In the end, there are few simple solutions to our energy quandaries, but the readings in this book may offer surprising new angles for considering the multifaceted significance of the "currents of the universal being."

Chapter Three

Grappling With Fossil Fuels and the Atomic Age

Audre Lorde, "Coal" (1976)

1934–1992

Writer and activist Audre Lorde, born in New York City to immigrants from the Caribbean island Grenada, once defined herself as a "black, lesbian, mother, warrior, poet." In her poetry and in her activism, Lorde argued that racism, sexism, homophobia, and discrimination against people with illness—Lorde spent the last fourteen years of her life battling breast cancer—are inextricably linked. She suggests that intolerance of difference is the common thread among all of these. In Ada Gay Griffin and Michelle Parkerson's documentary *A Litany for Survival: The Life and Work of Audre Lorde*, Lorde says, "Let me tell you first about what it was like being a black woman poet in the sixties. It meant being invisible. It meant being really invisible. It meant being doubly invisible as a black feminist woman and it meant being triply invisible as a black lesbian and feminist."

Discussion Generators

In her poem "Coal," published in 1976, Lorde plays upon blackness, both as it applies to coal from the ground and to her own identity. What meanings, both literal and symbolic, does coal have in our society today? Is it best considered a vital source of heat and electricity? A means of achieving energy independence? Is it a dirty energy source—one that should be phased out? Is it a cause of global climate change and therefore a threat to our survival? How might a poem titled "Coal" have a different meaning if it were written, for instance, from the perspective of a Kentucky coal miner? Is it possible to hold all these potential meanings in our minds at once?

I
is the total black, being spoken
from the earth's inside.
There are many kinds of open
how a diamond comes into a knot of flame
how sound comes into a word, coloured
by who pays what for speaking.

Some words are open like a diamond
on glass windows
singing out within the passing crash of sun
Then there are words like stapled wagers
in a perforated book—buy and sign and tear apart—
and come whatever wills all chances
the stub remains
an ill-pulled tooth with a ragged edge.
Some words live in my throat
breeding like adders. Others know sun
seeking like gypsies over my tongue
to explode through my lips
like young sparrows bursting from shell.
Some words
bedevil me.

Love is a word, another kind of open.
As the diamond comes into a knot of flame
I am Black because I come from the earth's inside
now take my word for jewel in the open light.

Muriel Rukeyser, "Time Hinder Not Me; His Arms Reach Here and There" (1957)

1913–1980

"Breathe-in experience, breathe-out poetry," wrote poet and political activist Muriel Rukeyser in her first book, *Theory of Flight* (1935). Rukeyser brought to her poems a deep moral sensibility, a keen sense of history, and a committed social consciousness—traits that made her one of the most important American poets of the twentieth century. Born into an affluent Jewish family in New York City, Rukeyser's early work was influenced by W. H. Auden in its intricate rhyming and regular meter. Later she wrote more freely, famously declaring in a 1968 poetic manifesto, "No more masks! No more mythologies!" In addition to her advocacy for the disadvantaged, Rukeyser covered a great range of interests, including science, in her writing, and in the 1960s and '70s became a favorite of feminists and the anti–Vietnam War movement.

Discussion Generators

In the poem sequence "Time Hinder Not Me; His Arms Reach Here and There," published in 1957, Rukeyser states, "I realize the consequences of that which was done on the desert, at Alamogordo," in reference to the Trinity Test, which occurred in the New Mexico desert in 1945. On the one hand, Rukeyser invites readers to understand the frightening potential of nuclear energy through the perspective of physics, in which human existence is just one small part of a series of cosmic cycles. On the other hand, she asks us to consider the preciousness—the sacredness—of an individual human life. How does this poem help us to understand both our immense potential and our moral responsibility?

I WILL NOT CARE FOR TIME, FORBIDDING ME

Peace the great meaning has not been defined.
When we say peace as a word, war
As a flare of fire leaps across our eyes.
We went to this school. Think war;
Cancel war, we were taught.
What is left is peace.
No, peace is not left, it is no canceling;
The fierce and human peace is our deep power
Born to us of wish and responsibility.

I realize the consequences of that which was done on the desert, at Alamogordo.

The work in the loss of mass.
The work in the lifetimes of the fixed stars.
The work in ideas of unstability:
divisible and transmutable as matter,
divisible and transmutable as idea,
The inner passage of lifetimes and of forms.
Relations of stars and of the stages of life.
The half-life of the forms.
The laws of growth and form.

I have tried to show the atom as a source
A source of energy.
I have touched on another question:
Might energy become a source of atoms?
If this relationship is real,
The universe passes along a way of cycles.
A process of matter dissolving in the stars,
Turned into radiation, passing through forms
Again to matter; again, perhaps, to birth.

Lilienthal said : I heard him saying words like these.
Almost like these words:
My convictions are not so much against, as for.
I believe—and I conceive law to rest here, as does religion—
The fundamental truth, the integrity of the individual;
That all we build be designed to promote and protect and defend
The integrity and the dignity of the individual.
This the essential meaning of our nation,
As it is essentially the meaning of religion.

Any forms, then, which make men means rather than ends,
Which exalt any institution above the importance of men,

Which rest on an arbitrary power over men,
Are contrary to that conception and my meaning.
That I deeply oppose, and I deeply disbelieve.
Out of this central core of a belief
That all men are the children of God
That their lives come first and are sacred,
A great belief grows in civil liberties,
In their protection; a repugnance to theft
Of these liberties and a human being's good name,
By innuendo or by open lies.
Here is no ethical standard, nor in the state
Which exercises blind powers over the human heart.
Occasionally, all these things are done
In the name of democracy.
They can tear our people apart.

I believe in the capacity of our central belief
To survive its trials provided only we
Practice, in daily life, daily, our truths.
They are affirmative. That is their hope in this world.
This I deeply believe.

Sandra Steingraber, "Shale Game" (2010)

b. 1959

Ecologist, author, and cancer survivor Sandra Steingraber is the author of several books, including *Living Downstream: An Ecologist's Personal Investigation of Cancer and the Environment* (1997) and *Having Faith: An Ecologist's Journey to Motherhood* (2001). Called "a poet with a knife" by *Sojourner* magazine, Steingraber has received many honors for her work as a science writer. The Sierra Club has heralded Steingraber as "the new Rachel Carson," and Carson's alma mater, Chatham College, selected Steingraber to receive its biennial Rachel Carson Leadership Award. She is widely recognized for her ability to serve as a two-way translator between scientists and activists, and she often incorporates her own experiences—particularly her battle with cancer—to personalize her activism.

Discussion Generators

The following article, from her regular column for *Orion* magazine, exemplifies Steingraber's talent for translating scientific concepts for a lay audience; she explains the process of "hydrofracking" as a means of extracting oil shale and why she believes this process would be bad for the ecology of the Adirondack Mountains. What metaphors does Steingraber use to make complicated scientific processes accessible to her audience? What other literary strategies does Steingraber employ, and how do they serve her overall purpose? Do you see any potential problems with using a rich, literary writing style to describe scientific topics? Does Steingraber's use of personal anecdotes undermine her authority as a scientist or does this mode of communication enhance her authority?

When I moved my family from a cabin in the woods outside of Ithaca, New York, into a house in a nearby village, it felt like a faith healing. I could walk again. A sidewalk stretched from my door out to a craggy maple tree and then connected with another sidewalk that headed down the block toward Main Street. Here was a track, upon which the wheels of a double stroller could roll, that linked me to coffee, library books, postage stamps, hardware displays, bank tellers, and a bus line. Hallelujah.

Out in the woods, foxes and newts had roamed our backyard, but I myself wasn't doing much roaming. The road that connected me and my children to the rest of the world was ditched on both sides and carried trucks and a 50 mph speed limit. Nobody was going to be tricycling along it, and trips to obtain cash, band-aids, or wallboard nails involved car-seat buckles, tantrums, and drive-through windows.

But now I sat on my front stoop and grinned. To be sure, the village sidewalks—century-old slabs of stone—were neither plumb nor true, but this was evidence that they had outlasted a generation of street trees whose roots must have lifted them and then, in dying, set them down uncrumbled but askew. Looking at the misalignments, I tried to guess where trees had stood in 1840. From a geologist neighbor, Bill Chaisson, I learned that our sidewalks are a form of shale—the mother of slate—created from marine sediments. That's when I noticed the marks of a vanished ocean on the walks' rippled surfaces.

And it is this vanished ocean—and a deeper layer of shale called the Marcellus—that has now placed the Finger Lakes region of New York, known for waterfalls, vineyards, and dairy farms, at the center of a looming epic battle over a new form of energy extraction known as high-volume slick water hydrofracturing. Or, to use the world's ugliest gerund: fracking. There are four stories to tell about it.

The geological story goes like this: Four hundred million years ago—before the Earth knew trees—the Acadian Mountains eroded into a nameless sea. Its silt sank into a trough in the ocean floor, together with the remains of mollusks, squids, and sea lilies. Under pressure, this graveyard turned into shale, forming a chalkboard the size of Florida. And the plankton and animals trapped inside became bubbles of methane. Because eroding mountains shed elements, this trough also captured uranium, mercury, arsenic, and lead. And so, in a bedrock layer that ranges from 2 to 200 feet thick, at a depth of 1 to 2 miles below the Earth's surface, at a temperature that ranges from 140 to 180 degrees Fahrenheit, extending for some 600 miles throughout West Virginia, Ohio, Pennsylvania, and New York, the shale's rock, methane, and heavy metals have remained locked together. Underlain by brine. Overlain by drinking-water aquifers.

Geologists refer to the Marcellus Shale as New York's ancient basement. Nevertheless, it comes blistering out of the ground in the little village of Marcellus—sixty miles and three finger lakes east of my village. That community became its namesake.

The engineering story goes like this: The Marcellus Shale holds the largest natural gas deposit in the United States. (What geologists call methane, energy companies call natural gas.) Drilling for gas by fracturing shale is an established practice, but, before the twenty-first century, capturing an effervescence of gas bubbles dispersed within a horizontal formation like the Marcellus was not profitable.

Enter slick water hydrofracking.

For this method, a drill bores down and then turns sideways. Explosives are detonated along the horizontal pipe, shattering the shale bedrock above and below. A pressurized slurry of water, sand, and chemicals goes down next. The water forces open the shattered rock, the sand grains keep it open, and the chemicals inhibit corrosion, kill algae, and reduce friction so that the released gas can flow up the pipe. Some of the water and chemicals forced into the fractured shale flows back up. And some of the water and chemicals—40 to 85 percent—stays in the ground.

A single fracking operation requires drill rigs, a compressor station, a network of pipelines, an access road, 2 to 8 million gallons of fresh water, 10 to 30 tons of chemicals, and about 1,000 tanker truckloads of water and toxic waste. About 4,000 wells are envisioned for my county alone.

The environmental story goes like this: In New York state, fracking represents the industrialization of a rural landscape and foodshed. If it goes forward, fracking will usher in the biggest ecological change since the original forests here were cleared. Road-building and pipe-laying will accelerate habitat fragmentation. Spills and seepage of toxic contaminants, including methane, into drinking-water supplies have been documented in other states and will certainly be an ever-present threat in the Finger Lakes region as well. Beyond this lie the unknowns.

The chemicals found in fracking fluid are unknowns both because their formulations are proprietary (Halliburton et al.) and because radioactive materials, heavy metals, and brine, freed at last from their subterranean chambers, combine with the chemicals in the flowback water. Where will it be treated? How will it be stored? We do know that fracking fluid contains benzene, a known carcinogen. Of the 300 other chemicals that are suspected ingredients of fracking fluid, 40 percent are endocrine disrupters and a third are suspected carcinogens.

The nature of government oversight is unknown because fracking is exempt from federal environmental regulations, including the Safe Drinking Water Act, the Clean Air Act, the Clean Water Act, and the Superfund law.

The impact on agriculture and public health is unknown because a cumulative impact assessment has not been done. Dust, noise, traffic, diesel emissions, ozone, soil compaction, light at night, methane plumes. How will these affect asthma rates, pollination systems, cancer risk, the growth rate of alfalfa?

There are also more elusive unknowns. Can fractures in the Marcel Shale radiate upwards? Could they connect with other passages, faults, fissures, and channels? Could they crack an aquifer? Can shattered bedrock safely contain toxic chemicals for 430 million years?

The human story goes like this: The Marcellus Shale could be worth a trillion dollars. It may provide enough natural gas to supply the nation's consumption for 2 years. Or 11 years. Or 20 years. Or 100 years. Leasing your land to a gas company can get you out of debt. It can allow you to retire.

Across the border in Pennsylvania, fracking is going full tilt, but, at this writing, there is a de facto moratorium in New York, as we await the release of a state review. Meanwhile, a pipeline has been laid from Corning to Rockland County, and millions of dollars are being spent quietly issuing leases. In my village, 14 percent of the land is already leased to gas companies. In the county, 40 percent. "The shale army has arrived," said a representative from an energy company. "Resistance is futile." And, indeed, in December 2009, ExxonMobil purchased a large natural gas company, a decision widely viewed as a game-changing commitment to fracking technology.

Nevertheless, at a recent meeting at my village firehouse, candidates for board and mayor declared their opposition to fracking. A public meeting about fracking at the village library included lively discussion about a community on nearby Keuka Lake that had turned away fracking wastewater trucked in from Pennsylvania. An older man in the audience declared passionately, "We have to be ready to lie down in front of the trucks." On the way home, walking on an unbroken sidewalk made of shale above an as-of-yet unshattered bedrock made of shale, my son said, "We shouldn't wreck this place down, right, Mom?" And his words drew a battle line across my heart.

Rebecca Solnit, "April Fool's Day," from *Savage Dreams: A Journey Into the Landscape Wars of the American West* (1994)

b. 1961

Award-winning San Francisco–based journalist, essayist, environmentalist, historian, and art critic Rebecca Solnit has spent much of her writing career examining the effects of energy policy on the American West, most notably in her 2004 book *Savage Dreams: A Journey Into the Landscape Wars of the American West*. Of her undergraduate work at San Francisco State University, Solnit has said, "The best part about the critical training I got in the visual arts is that it was really just about reading things carefully and asking questions about meaning. The subject could be an artwork, but it could also be the history of nuclear physics or national parks or the representation of Native Americans or the perceptual and spatial changes the railroad brought." The wide range of subjects on which Solnit has written—from walking to nuclear energy, geology to art history—is testimony to her aptitude for applying serious, rigorous thinking to interdisciplinary subjects.

Discussion Generators

Many people, including serious environmentalists, have suggested nuclear energy as the best way to meet the anticipated energy needs of the twenty-first century. It's less expensive than many other forms of energy, it doesn't contribute to anthropogenic climate change, and it isn't likely to be subject to shortages in the foreseeable future. Perhaps the biggest as-yet-unsolved problem associated with nuclear energy has to do with the storage of nuclear waste. In the excerpt from *Savage Dreams* that follows, Solnit critiques the U.S. government's efforts to deposit spent nuclear fuel at Yucca Mountain in the Nevada desert north of Las Vegas, which seems to be off the table at this time. Do you find Solnit's case against the Yucca Mountain site compelling? If so, what criteria should be used to determine where nuclear waste is stored? Is the risk too great to continue using nuclear energy at all?

Just as the MX Missile was giving up the ghost in the early eighties, Yucca Mountain succeeded it as a doomsday future for the state of Nevada. The government, which hasn't been able to make any conventional use of public, or Shoshone, land in Nevada, seems hell-bent on making it useless for everyone and everything for all time.

There always has been one problem with nuclear power, even had it worked the way many scientists envisioned it working in the 1950s and a few still dream it could nowadays. It was supposed to be an open-ended source of energy, cheap and without the output of pollutants coal- and oil-burning plants had. But it used quantities of a dangerous metal, uranium, and transmuted it into the most toxic substance on earth. The plutonium and other wastes weren't considered pollutants, since they didn't go into the environment—but where would they go?

Now, nearly half a century into the nuclear age, no one has answered that question, and as long as it remains unanswered, the nuclear power industry is in jeopardy. Much of the nuclear waste produced in weapons production has been unsafely stored—including rusting barrels in the ocean off the San Francisco coast and leaking storage tanks in Hanford, Washington, time-bomb monuments to the underestimates of the past. Spent nuclear fuel has been accumulating in the cooling tanks around nuclear power plants, where the operators are waiting for the government to make good on its promise to take this stuff off their hands. Spent is an ambiguous word: Although the uranium has outlived its usefulness for generating the subcritical reactions that fuel a nuclear power plant, it has partially mutated into more dangerous things—into strontium, cesium, and plutonium, and it is still very hot literally and radioactively. The half-life of the first two elements is short; but the half-life of plutonium is 24,000 years—and half-life means the amount of time it takes half of the element to decay into other elements, so it takes several hundred thousand years for plutonium to decay enough. As far as many scientists and activists are concerned, adequate storage is an idea that has not been realized yet, and may be unrealizable.

In 1982 Congress passed the Nuclear Waste Policy Act to initiate a quest for a long-term storage site for the rapidly growing temporary storage piles of nuclear waste. The initial plan was to have a site in the East and a site in the West, but several eastern states quickly shrugged off the duty. Three sites—one in Texas, one in Washington, and one in Nevada—were chosen, and political pressure eliminated two of the places, so in 1987 Congress passed what gets called the "Screw Nevada" bill, which prematurely settled on Yucca Mountain as the only site to be studied. Since then the state of Nevada has worked hard to keep the waste repository out, even though its senators and local politicians tend to favor nuclear testing (for the jobs, they say). The plan is to store 70,000 metric tons of high-level nuclear waste in the mountain, where the DOE has decided it must be isolated for 10,000 years to be considered safe (though the elements will be hazardous many times longer).

When I asked about Yucca Mountain, Bob Fulkerson sent me to talk to Steve Frishman, a geologist working for the state to prevent the waste repository from opening. Frishman, who worked himself out of a job when Texas defeated the proposal to site a repository there, told me that most of the country's commercial nuclear waste is produced in the population-heavy, resource-light East, and most of the plans for storing it point west. "In 1989 the state Senate passed a law that says in very few words that storage of high-level waste in the state of Nevada is prohibited. That's the major states-rights question that's going to be tested if the permitting process goes forward. It's maybe as significant a constitutional crisis as the Civil War, and Westerners are becoming more and more conscious of the fact that the East is eyeballing the West as a major dump site—for everything."

I asked him about what Bob had told me during that seminar under the sky at the Nevada Test Site: that the waste slated for Yucca Mountain is so literally hot it will boil for a thousand years. I had never heard anything like it besides medieval descriptions of the torments of hell. Frishman told me that Bob had underestimated both the temperature and the time. One of the plans for keeping water from reaching the waste, he said, is to place the spent fuel rods close enough together to keep the storage area above boiling temperature for millennia. This would succeed in vaporizing any water that might approach the area—but the rocks around would fracture from the heat, creating other problems.

I'd gone to visit Frishman in his office in Carson City a month or so after a major earthquake did a million dollars worth of damage to the DOE study buildings around Yucca Mountain, and he spent a whole morning telling me about geology, politics, and radiation in a Texas drawl, chainsmoking, and cackling at the madness of the plans. The main assumption made by the Department of Energy geologists is that Yucca Mountain is an ideal repository because of its low water table, and the main objection by its opponents is that the water table there is an unstable, even a mysterious thing. There are earthquake faults running all across the Test Site region too, and there has been volcanic activity in the area. Frishman said, "There's certainly a possibility of recurrence of vulcanism. When you drive along 95 the Lathrop Wells cone is pretty evident. As a matter of fact, the first time I was driving here from Texas, I was driving along 95, didn't know I was driving by Yucca

Mountain or anywhere near it, I knew I was in that vicinity but I didn't even know that you could see it from the road, but I just looked over and said 'Man, there's a young volcanic cone—what's going on around here?' It turns out that the most recent activity associated with the Lathrop Wells cone could have been as recent as five to fifteen thousand years ago, which means it's very recent." And he told me that any volcanic or seismic activity could change the water table.

I asked him, too, why the DOE had chosen 10,000 years as the timespan for isolating nuclear waste, and he really hit his stride. "They did that for two different reasons. One is that in 10,000 years the level of radioactivity gets back to about the level of radioactivity of a fairly rich uranium ore, so it goes back to something that is no more radioactive than the most radioactive thing that someone would come into contact with in normal life. That was sort of an afterthought. The other thing that was the real driving factor is that if you're going to do a probabilistic risk analysis and you're looking out to some distance into the future, when you get out to about 10 to the 4th, meaning 10,000 years, you start vastly increasing uncertainty just because of our inability to predict geology and human futures and so on, our inability to predict climate. . . .

"Lake Lahontan was a significant lake for ten, fifteen thousand years. It's been drying up for the last 10,000 years or so; we're finding more and more evidence that it's been a cyclic drought. Most of the last 10,000 years has been very much wetter than now. We really don't know how to predict climate, but we have to assume that we're in an interglacial and it's going to get wetter again sometime, and sometime probably starting within the next 10,000 years. So you have to assume more rainfall. If you assume more rainfall, you have to assume more recharge to ground water, you have to assume some type of a rise in the water table. The DOE people have looked at it in terms of if we went back to the way it was 15,000 years ago how much would the water table rise? And in their very simplistic calculations you can see it rise about 400 feet, and that's about halfway from where it is now to the proposed depository horizon. The question is as simplistic as their calculation is, and the assumptions about infiltration; if they say it can go fifty percent what if they're 100 percent wrong, and the water's all the way up there?"

Another geologist, Charles Archambeau, told the *New York Times*, "You flood that thing, and you could blow the top off the mountain. At the very least, the radioactive material would go into the ground water and spread to Death Valley, where there are hot springs all over the place, constantly bringing water up from great depths. It would be picked up by the birds, the animals, the plant life. It would start creeping out of Death Valley. You couldn't stop it. That's the nightmare. It could slowly spread to the whole biosphere. If you want to envision the end of the world, that's it."

Drum Hadley, from "A Giddy Up and a Whoa and a Catskinner" an excerpt of "A Jack of All Trades" (2007)

b. 1938

Never quite comfortable with any of the labels that have been applied to him as a writer—Beat poet, cowboy poet, nature poet—Drum Hadley has always been most at home doing the daily work on his ranch near the place where Mexico, Arizona, and New Mexico meet. For decades he has been an activist for ecologically sustainable ranching practices, and his poetic vision often expresses this outlook. In poetry collections such as *Voice of the Borderlands* (2007) and *The Light Before Dawn* (2011), Hadley depicts life in the American West as heard through campfire yarns and picaresque vignettes. Hadley characterizes his poems as "practical in the sense of getting a job done like saddling a horse or a bronc," and the rough-and-ready voices of his narrative personae lend his poems a unique authenticity.

Discussion Generators

In this excerpt from the poem "A Giddy Up and a Whoa and a Catskinner," published in his 2007 collection *Voice of the Borderlands*, Hadley examines the uneasy relationship between working people and the machines on which they depend to carry out their work. Love them or hate them, machines—Hadley seems to be saying—are the currency of our lives. What does Hadley's use of vernacular speech in this poem tell us about the relationship between working-class people and machines? How might the meaning of the poem be different if Hadley had not used language in this way? Are there ways in which working-class people relate to machines differently from middle- and upper-class people? How might these differences influence the way people perceive the fuel that powers these machines?

> "Goddam, how the hell are you?" the catskinner yells,
> Roaring with his one-ton truck, off the hill and into the canyon.
> "The brakes went out, almost went over a cliff.
> Air compressor tryin' to kill my ass.
> Fuel trailer tryin' to kill my ass.
> Hell, I'll be buried under one of these blackjack oaks.
> Paid one hundred seventy-five dollars
> For a brand new generator for the bulldozer,
> But the son-of-a-bitchin' generator don't work.
> Fixin' one is worse than a goddamn chicken,
> Pickin' his teeth with a paper toothpick.
> Goddamn sorry-assed piece of shit,
> If anything kills me, it's gonna be that heap of scrap metal.
> Hell, you trust the hydraulics to hold up her blade,
> She'll cover your ass in a heartbeat.
> Havin' a bulldozer is about as close to havin' a wife as you can get.
> Spend every last cent you got on her,
> Eats like hell, keeps you exhausted all the time,
> Breaks you on the fuel bill, and runs around hotter'n hell all day.
> Let's shut this son of a bitch down and take a break."

Rick Bass, from "A Short History of Montana," *The Heart of the Monster* (2010)

b. 1958

Rick Bass is best known for the ardent environmental activism expressed in books such as *The Ninemile Wolves* (2003), *The Book of Yaak* (1997), and *The Lost Grizzlies* (1997). In 1987, he moved to the remote Yaak Valley near Troy, Montana, where he has worked tirelessly to protect his adopted home from roads and logging. It may surprise some readers to learn that Bass began his career as a petroleum geologist, a contractor for the same oil conglomerates who come in for harsh criticism in his 2004 book *Caribou*

Rising: Defending the Porcupine Herd, Gwich-'in Culture, and the Arctic National Wildlife Refuge. Bass's personal experiences on both sides of the fossil fuel extraction debate—first as oil man, then as activist—give his writings a special authority. The selection reprinted here comes from the 2010 book *The Heart of the Monster*, which he coauthored with Montana writer David James Duncan, in opposition to the mining of the Alberta tar sands in Canada and the transportation of mining equipment through the Pacific Northwest and the Northern Rockies en route to Alberta.

Discussion Generators

Alluding to his own background as a hunter and petroleum geologist, Bass argues, "It is the destiny and identity of each of us to be a taker"—meaning, perhaps, that there is something uniquely human about extracting natural resources from the earth and putting them to our own use. What do you make of Bass's assessment of human nature? How can we account for his passionate criticism of Montana's politicians and his fierce opposition to the Alberta tar sands project? Why does Bass begin this section with a discussion of a diesel truck that has gone off the road—how does it illustrate the problems, as Bass sees them, associated with tar sands extraction?

Another truck has gone off the road, up in Canada. A regular-sized tractor-trailer carrying diesel has slid off the road on U.S. Highway 12 in Idaho, spilling fuel into the Lochsa and jamming up travel for several hours. A few more hunting and fishing clubs have signed on to the protest, and a few hundred thousand more people who skim the newspapers have noted and filed away in their busy minds. *Oh, another lie, the corporations were mistaken*, and—perhaps they note this—*the governor, entertaining though he is, was mistaken, was not in control.*

This is not a familiar cognitive association: people are used to being confident that the governor is always on the right side, and is always "for the little man." Imperial will come in later and spread some public relations salve over this region, like roadworkers applying patchwork asphalt to a pothole—the twenty or so advertisements sociologists tell us are all that is required in the deadly alchemy of our times to spin lies into brief truth. You have to hear a lie twenty times before it becomes a truth. I find this amazing.

But a few will remember that spill. Our lives and our minds are filled to the brim and often to survive we all decide to hear only what we want to hear: but like it or not, already, the trucks are sliding, and it's not even winter.

Up in Canada, still another truck has gone off a road up there, in the rolling hills of the region formerly known as the Great North American Boreal Forest. Again, no ice, just a winding road and that pesky human error. The Royal Canadian Mounted Police, who investigated the spill of the Mammoet equipment at milepost 61 on Bretonne Road, reported that the driver received a broken leg and was cited, and that the accident had been investigated. Traffic there was blocked for only a couple of days, because there was a crane nearby that was capable of clearing the obstruction, and commerce, and the world's work, was able to proceed.

Still, and already, the rivets are beginning to pop at the seams. Imperial's too big to fail, but might the project be too large, too lunatic, for even the largest company in the world? The tar sands of Alberta are not all that dissimilar from the asphaltic deposits around and beneath Los Angeles, where perfectly-preserved Ice-Age mammals—saber-toothed tigers, North American rhinos and camels from 30,000 years ago—lie trapped and forever anguished in the bubbling seepages of La Brea. Activists report that Imperial drafted plans to excavate there, too—that Imperial is taking hundred-year leases on the ground beneath the city, and will excavate there when Alberta is finished—though there is no nearby water with which to flush and steam the tar of Los Angeles, other than the ocean itself.

There are tar sands in Utah, as well. After killing Indians we will move on to the Mormons. Instead of flushing the tar with the Athabasca, we can use the Snake River itself.

What seems unthinkable now, they know, we may well be clamoring for, insisting upon, and soon. As the world burns further—the crackling of flames so loud it makes it hard to sleep, and even the nights so hot that we must run our air conditioners ceaselessly, to try to mute the burning—we will be asking for, demanding, more. Whether a governor in some obscure Western state was once upon a time called Governor Oil or not really won't matter to anyone, everything will have burned or sunk back down into the earth anyway, there will only be fire, and sky.

The trucks are failing already. Montana is the weak link in the supply chain of this fevered dream and the trucks—improbably as elephants laboring to cross the Alps in snow—are failing already, even without the burden of their mammoth loads. I wonder if the governor ever wakes up troubled, understanding that for the first time he is exposed to the vulnerability of having made the terribly wrong choice, and considers an exit, an amendment: if he ever considers turning away from Imperial's power and going back toward his own self-generated power, his populism, his purer burning.

The governor is on the wrong side. He has his spokespeople saying that this issue isn't a two-by-four of a problem, but a toothpick.

Think about it—how problematic can a toothpick be? It's just a toothpick. He says everything will be fine. He says the big oil companies who asked him to make the corridor permanent didn't really mean it, that they don't really want to be permanent after all. He tells us to relax and forget we heard that. Everything is under control.

For now, Imperial has his back, is pumping money into the wound of his ambition like Congress pumps money into automakers, or banks, or whathave-you. Right now however—before the money begins to move—is our brief uncontested shot, our one chance to tell the story of what Montana is like, and what our lives are like, here in the garden, before the spell is cast over us.

We will know they—the devourers—have arrived, coming as they will in a first-wave advance of trucks like a plume of poison spreading loosely and slowly toward us, when the television, radio and print ads, and *faux*-grassroots agitprop, begins, the rhetoric about the tar sands' mining being necessary so that we can stop having our soldiers killed in Afghanistan: a murderously disconnected piece of political usury. If the governor wants us out of Afghanistan then he should work to that effect. Killing Indians in Canada and checking *Yes* on the box that asks if we want global warming to be irreversible, with its trillions of dollars of damage and untold hundreds of thousands of lives lost, is not a matter of choosing to fight in Afghanistan or not. Hooking us ever-deeper on Canadian oil and a petro-economy increases the degree and duration of our relationships in the Middle East, no matter how much oil we steam from the great pit in the north.

We will know that we in Montana—the weak link in the chain, the one best place we can battle to a standstill the monster of Imperial, and monster of our own procrastination—are winning when they begin to change their names—finally, no longer arrogant; finally, knowing (and for the first time) just the least little bit of concern. The modules' transport ship, the *Exxon Dong Bang*, will change its name to the Hope or the Patriot, and Imperial itself will spin off and splinter into various mercurial divestitures, each as temporal as a mayfly. They will form non-profits for Indian health care even as they are killing the Indians, killing everyone, and will keep pumping money into these new divestitures, more difficult to track—much less corner and capture—than any wild-spooked elk herd thundering in twenty directions down off the mountain and into the thick ravine below. It will take a sturdy and willing heart to follow them and find them.

Imperial will continue to pump money into these divestitures, and into their secret politicians, redistributing their power as if through giant cables filled with shimmering electricity. They will change the names of their companies to words like ConAgra or Altria or Nutritia or Sweetia. They will ask us to accept them. They will negotiate with us—will tell us that they will accept and feed us if we will accept and feed them.

They are shoving us around like they own us. They are naming themselves when really, though we never exercise this power, it is we who possess—again, through our charters—the power to name them.

You may operate in our state, we can tell them, *though you must do so under the name of Adolph Hitler, Satania, or Putrid, Inc.*

We know what's coming. Right here, right now, is our first and last uncontested shot. They had a two-year jump on us, in those secret meetings with the governor, but now is our time, in these coming weeks. They have been planning this for a lot longer than two years, and I think they were surprised when in 2004, a Democratic governor won in Montana, in the reddest state in the country. But in time, it came to be a pleasant surprise for them. A lap dog, some would say, but what a politician! Good enough to carry wildly a wildly red state.

Needless to say this is not a red or blue issue. Like war, or a garden, life is not red or blue.

All my life I have leaned heavily, hungrily, on what's real. I've reached down into the ground, in Mississippi, Alabama, Louisiana, and Texas, and have pulled up oil from three thousand feet below, cracking into the brittle sand reservoirs of old oceans-turned-to-stone. I have reached into the forest, into the mountains, and summoned—extracted—bull elk, as well as mule deer bucks and whitetails. It is the destiny and identity of each of us to be a taker—no living thing can exist independent of this real and made world—but being human, we are perhaps uniquely gifted with the compensatory mechanism, the ability to give back, or to try to give back, in some accordance with our taking.

To know the price of our existence is the foundation, I think, of all morality and religion. Ultimately, such knowledge is a path to peace, if any is to be had.

I used to think that when the world finally changed for the better, the central pivot-point for that long-awaited slow turn of history would somehow be visible as a vertical structure, like a landmark rising from the plains, visible from hundreds of miles away: a real and positive space on the landscape—not just culturally or imaginatively, but a real and symbolic thing—toward which we could all travel, committed to make a change, to turn a corner and pass beyond.

How strange it is for me to consider that the massive landmark—the point around which we must all pivot, and now, if we are ever to change—does not rise above the horizon at all, but is a gaping pit, a negative space, and one which, despite being the largest in the world, isn't visible until you're right up on it, standing on the edge.

The Alberta tar sands are such a threat to life as we know it on this green earth that the damage there will make the BP oil spill in the Gulf of Mexico look like a broken finger nail, a sneeze, a modest cough, a hiccup, a blink. Further development of the Alberta tar sands will by comparison make the BP spill in the Gulf look like a shot of multivitamins, good for the earth. When pressed on this point, the governor and his spokespeople dodge the issue that they could stop this, and with the malice of a drug pusher rather than the courage and foresight of leadership, taunt us, saying, *Don't put gas in your cars if you don't like it.*

My governor, what word, what words, could awaken you from this dream? It is as if you have fallen asleep, or as if a hood has been placed over your head. What falconer owns you, now, and can you not uncover it yet, and return to who you were, and where you came from? This life is such a short one but you still have time.

The shape and the measure of things is getting beyond us. The list of things that are slipping from our control—yet for which we are accountable—is growing, to the point where we are going to need something larger than our own selves to get us out of this jackpot.

Every bent and twisted fish, every sick frog, every vanished bird species, every dying Indian, is on our increasingly perverse wish-list, as we dodge and weave, trying to avoid looking at the real cost of tar oil—and as we grow still and quiet and humbled and shamed by our affluence as well as our inaction, we are coming slowly to realize how very much we could use a little magic.

We sure don't deserve any, but that's part of what makes it magic. Some people call it grace, no matter; I'm talking about the same thing.

And it is close. It is close. It is so close that if you listen you can see it, and you can hear its breath: as when you are in the woods on a cold morning, and are sitting very still, just listening. It is close.

All we have to do is have the courage to speak our name. The one thing given to us when we came into this world, and the one thing Imperial and Exxon cannot take from us. Not even Imperial can take that. All we have to do is write our name on a sheet of paper—pass a citizen's initiative—and they go away.

I get weary of words. They are useful—they are the building blocks of democracy, and of revolution—but without action, they are nothing more than the relicts of ancient civilizations, crumbling across time. They can be the foundations of history but without action, they are—even in the moment of their first-being-created—like mansions uninhabited by any residents, or like a man or woman uninhabited by a soul. They need the spark of action, they need the breath of the living.

I am getting weary of them, important though this short history is, and this issue. Imperial and Exxon may seem patient, but they aren't. The shark must keep swimming, must keep eating. Not a day may pass without the largest possible bite. They are a ghost, a nightmare, are becoming a phenomenon unto themselves, like the ice age, like the northern lights, like a volcano, an asteroid.

Imperial is not at all about words, only action. Now they have brought more modules up the Snake to the so-called inland flat-port of Lewiston, in defiance of the court cases that are still to be decided. As if knowing already the outcome—Imperial's success, and our demise. As if so accustomed are they to always getting what they want, rolling over everything, that they cannot and will not pause for anything—not public opinion, and not rule of law, nor court of man. They are busy, after all, they have oil to deliver to a waiting public. They are not patient. The refinery-tubes sit down in the pit, the dark canyon, of Lewiston, in the eight-foot shallows of the dying Snake River, where there has been discussion of laying a set of railroad tracks down the center-spine of this once-great river—which can be great again—so that after the water has dried up Imperial can continue to truck the tar sand equipment northward in that fashion.

What few beleaguered salmon remain in the Snake will be writhing in the mud-cracked shallows. Eagles will feed upon them and Indians and non-Indians alike will be able to walk down to the dead river's edge and pitchfork the last of them into wicker baskets as was done in the old days.

It would be tempting to say the salmon won't ever be coming back. But nothing, really, is forever. Ten thousand years from now, the ice itself might come back. A hundred thousand years beyond that, it might melt again, and gigantic ocean-going silver-sided fish might tentatively explore the ice-carved canyons—gliding over the sutured lattice of those tracks, and the sunken carapaces of barges and refineries.

Along the Rocky Mountain Front, shaggy ice bears, weighing in excess of a thousand pounds, might glow burnt umber in the morning sun. Who knows what world lies beyond this one? We are each alive for only a few more days, hours, years; is it not richer to live these last hours with courage and resolve rather than under the brown-nose kiss-ass sell-out of being owned by Imperial?

William Heyen, "Pterodactyl Rose" (1991)

b. 1940

In a 2001 interview, William Heyen said of poetry, "I still believe that only poetry can save the human world. But I'm not talking about our word constructs, these lyric and other poems that we read and write, but about a poetic conception of our place here, of the earth as One, of thought as integration until we realize we must change or we will die out." Whether he writes about the Holocaust or the science of ecology, this sense of urgency, of seeking to stave off extinction, infuses much of Heyen's poetry. His books include *Noise in the Trees* (1974), *Long Island Light* (1979), *Erika: Poems of the Holocaust* (1984), *Pterodactyl Rose* (1991), *Crazy Horse in Stillness* (1995), *Pig Notes & Dumb Music: Prose on Poetry* (1998), *Diana, Charles, & the Queen* (1998), and *Shoah Train* (2003), which was a finalist for the National Book Award for Poetry.

Discussion Generators

In the following poem, Heyen wrestles with the paradox of his own use of gasoline—the cognitive dissonance he experiences when he remembers that it is composed of the bones of organisms now vanished from the earth. When the poet looks into his rearview mirror, what sort of past—and future—does he seem to envision? What does the poet evoke through the words "behind me where I'm going"? When the poet says "like you" in the first stanza and "unlike you" in the second, what sort of reader is he addressing, and what sort of relationship does he imagine between poet and reader? What response does Heyen hope to elicit from his readers?

Like you I drive my ten
thousand American miles a year
burning fossil fuels (conversion
to a ton or two of carbon)

but maybe unlike you I peer
into my rear-view mirror imagining air
filling with insects & plants maybe
Triassic dinosaurs

turtles Devonian dragonflies & lilies
such beak & leaf & wing & vine
profusion the past assuming
extinction's shape behind me where I'm going

wild with this prayer of mine,
& longing.

Ricardo L. García, "Sign at the Mine: No Women or Bears Allowed," from *Coal Camp Days* (2001)

b. 1940

On his faculty page on the website for the University of Nebraska–Lincoln, where he is currently a professor of education, Ricardo L. García writes that "education should be a liberating force in the lives of people." The same guiding philosophy could be applied to García's fiction, which represents ordinary people as they struggle to make a living and raise families, to educate themselves despite obstacles, and to pursue the American Dream. In his 2001 novel, *Coal Camp Days*, García presents a vivid depiction of life in a New Mexico mining town, often incorporating folklore and stories told by the town's many other ethnic groups, among them Italians, Slavics, Greeks, Latinos, and African Americans.

Discussion Generators

In this excerpt from *Coal Camp Days*, García considers the role of family within the larger social and environmental contexts of a community whose economic roots are in the coal mining industry. Why does Matías want to work as a coal miner? How does Mr. Malcolm's story about shaking hands with the bear reinforce Matías's romantic ideas about coal mining? The sign at the entrance to the coal mine says, "NO WOMEN OR BEARS ALLOWED," suggesting a relationship between the ways women and bears are excluded from the mines. Does the sign provide any insight into why Matías's mother does not want her son to become a coal miner like his father?

My claim to fame was toppled by a real hero known for brave and daring deeds. Our trout fishing expert and neighbor, Mr. Dick Malcolm, was the first coal miner to serve as fire boss in the Chicorico mine. He cast the die for being a good fire boss. A fire boss was highly respected by the miners. His job was to inspect the mine before each shift to make sure the mine was safe to work. In the Chicorico mine, the fire boss had to be especially careful about damp-gas and faults in the mine's bedrock ceiling.

Damp-gas was the methane released when coal was dug or blasted, and carried in the dust that settled on the mine's floor and walls. When it accumulated, explosions were possible. Damp-gas was easily stifled by simply sprinkling water on the dusty walls. Faults in the bedrock and coal seam were harder to detect. The fire boss would tap the ceiling's roof with the end of a pick handle, listening for hollow spots indicating a break or fault. Faults were buttressed by using log props to support the ceiling.

The miners selected a conscientious veteran who knew the craft of mining and could detect unsafe working conditions. The stories about Mr. Malcolm were legion. After the union was organized, he was the first fire boss to be appointed by the miners. No hazard got past him, be it poor wiring, weakened cross-ties, a possible faulty bedrock, or gassy conditions. He was tough on hazards.

During supper one Saturday, Juan anxiously awaited to tell about one of Mr. Malcolm's latest feats. To Juan, it was bigger than life and took more than a grain of salt to digest. Dad made Juan wait a bit before telling the tale. For Dad and Mom, supper was a crucial gathering of the family where serious concerns were discussed, with no petty bickering allowed. All topics were allowed. Important issues or family matters received the greatest attention. Yet suppertime talk was always convivial—the time and place each day we could be heard by the whole family.

Just before supper, Dad would ask us to sit with him in the living room to listen to Gabriel Heater, a radio news commentator. We concentrated on what the news commentator reported because Dad would test us as we ate. At supper after the blessing, the news was discussed. Most of the news was about the European and Pacific war efforts. We learned about geography, different peoples, and cultures. We also learned much when the commentator talked about the war's economic and financial underpinnings, especially about war bonds and other emergency means for funding the war.

"José, these are bad times in Europe. Tell us about it." Dad led the discussion.

"We're on the beach in France. The biggest invasion of the war."

"How do you know?"

"Gabriel Heater said that! From the Armed Forces Radio News."

"We're having it the worst," Juan chimed in, "at Omaha and Utah Beach."

"The English had an easier time at those other beaches," José injected.

"Gold, Juno, and Sword," Angela recalled the names of the other beaches. "Their invasion was much easier. They're calling the invasion 'D-Day.' It's a tremendous attack—the largest ever. Soldiers from England, Canada, Australia, America. Our soldiers have run into fierce resistance by the Germans."

"And what about the paratroopers?" Dad checked our memory.

"Oh! They went over first." Angela had a sharp memory for details. "Dad, is it true, that Arturo flew the paratroopers for D-Day?"

"*¿Quién sabe?*" Dad sighed. "We don't know where he is."

"That's why we think he's there," Mom sounded worried. "Y, tambíen Edward Heard. He wrote his mother. Mrs. Heard could barely read the letter. There were many holes in it, where words were cut out."

"How come?"

"Because we're at war with Germany and Japan," Dad quickly answered. "Spies might get the letter and find out about our troops."

"Arturo's service flag finally came. I'll hang it after supper." Mom held up the four-by-six-inch service flag, with a blue border, a red inside border, a blue background, and a white star. The single star signified we were a one-star family with one person in the armed services. A gold star signified the person had died. For some reason, we had just received the flag although Arturo had been in the navy for a long time.

"Where will you hang it?"

"In the living room window."

"Good." Dad stroked the flag fondly, rubbing his fingertips across the weave of the star. "Now we can show all who pass by the house, we have a son in the war."

Juan was more interested in the here and now. He could hardly wait for the serious discussion to end. Whenever he was nervous or impatient, he would tap rhythm with his left foot under his chair. I sat next to him and could feel the vibrations. He was aching to tell about Mr. Malcolm.

"I hope everyone's done?" Juan pretended to be calm. He didn't want to upset Mom or Dad by appearing impatient. Then he'd never tell his story.

"What's the rush, Juan?" Angela quizzed, "I have new duties at the store. I now work at the cash register as well as assist Mr. Ruker, the manager."

"Clara?" Dad glanced at Mom, wondering who would help Mom in the house. This was the first he had heard that Angela worked in the Company store.

"O, sí, Manuel. I told Angela to find work. She doesn't like working in the house. Anyway, I have the boys to help."

"And Ramona?" I chirped. Ramona hooked her little finger in the side of her mouth, smiling.

Mom jumped to Ramona's defense: "O, no! She's too little."

Mom's comment about Ramona made me grin, causing me to speak without thinking, "Ramona has no chores."

Mom immediately read my thoughts; she knew I was jealous of Ramona. "Don't worry, her time will come." Ramona kept smiling; I'm not sure she understood my jealous motives, but she knew we were talking about her.

Dad ignored me, turning to Angela. "And, what do you do at the store?"

"Many things—help with shelving, help with inventory. I'm a cashier as well."

"So you handle people's money?" Dad frowned.

"Yes. And their scrip . . . when they use it."

"Do you keep some of the money for yourself?" I insinuated, meaning to joke.

"Certainly not! How dare you say that!" Angela snapped defiantly, glaring at me. Integrity was no joking matter for Angela.

"Honey," Mom interrupted, "don't be too harsh on Matías."

"Well, I'm honest. I would never shortchange a customer—not on purpose. Mr. Ruker trusts me so much, he may have me keep the books."

"That's good," Mom softly commented. "Mr. Ruker wouldn't say that if he didn't trust you."

"My word is my bond, Mother. I pride myself. I can be trusted to handle money."

"Yes! Yes!" Mom was a calming influence, "I'm proud of you. If you need help, your father keeps the books for the union. He can show you."

"That's okay!" Angela cheered up. "I've been watching Dad, and the Sisters at St. Patrick's are teaching a class in bookkeeping." Angela was a good student. The nuns liked her a lot.

"I'm sorry—just joking." That's all I could say. I felt bad, but I meant no harm. I really liked Angela. She was nice to me. I was the first child Mom had asked Angela to help with. She'd even changed my diapers. And, when Ramona was born, Angela assisted Mom by serving as my full-time nanny. As Mom grew stronger, and no longer had to stay in bed after Ramona's birth, Angela still helped with my care and really made me feel special.

"Try not to joke so much, Matías," Angela reminded. "Sometimes jokes hurt people."

Juan poked me in the ribs. By now, his left foot was really tapping! He was anxious to tell his story. By horsing around, I was taking precious time from Juan.

"Did you hear about Mr. Malcolm?" Juan butted in. "He shook hands with a bear!"

"N-a-a-ah! Huh?" José and I exclaimed in disbelief.

"Yes! Yes! This morning, he finally did it." Juan's face was beaming with a smile from ear to ear. He really wanted to spin his tale, but he had to wait for an okay from Dad.

"Did what?" Dad gave Juan the opening he needed. Now he could tell his tale.

"Yesterday this bear was looking for chokecherries in the bushes. That's what Mr. Malcolm told me. He spotted it when he went to work early; he was fire boss for the day shift. He said the sow was alone without any cubs. Anyway, she found some chokecherry bushes by the mine entry. He never paid her no mind; he went into the mine and inspected it. When he gave the okay, the men came in the mine to work. He noticed the bear was up the slope above the mine entry, eating some chokecherries. He went back into the mine to start working.

"Mr. Malcolm figured after she ate all the chokecherries off those bushes, she smelled more chokecherries right inside the mine. She followed her nose. Bears can smell pretty good. They have good noses. Her nose took her right up to the mine entry. The shaft was posted:

NO WOMEN

OR

BEARS ALLOWED!!!

"The miners say women and bears bring bad luck in the mine. Women bring the worst luck. But bears are plenty bad too. Anyway, she couldn't read the sign. So she did as she pleased . . . went right into the mine. Inside, she found some of the miners' lunch buckets. They all smelled yummy. She liked the lunches that had chokecherry jelly sandwiches in them. With both paws, she tore into the lunch buckets. They didn't stand a chance. She ripped them open and ate everything down to the last crumb, 'specially the chokecherry jelly sandwiches, that's what Mr. Malcolm said. . . ."

Juan paused for dramatic effect, waiting for our thoughts to catch up to his story. He paused long enough for the tension to mount before continuing. "She was still hungry. So she walks into the mine where some of the miners were working. *¡Hijo mano!* Man alive! Did they get scared!

" 'Oh, brother!' one miner said. 'A bear in the mine is bad luck. A woman in the mine is worse luck. But a woman bear in the mine? Double-bad luck!"

"The miners were so scared they wouldn't do any work. The bear might eat one of them. Or, two of them!"

Another pause by Juan . . . "But, Mr. Malcolm wasn't scared. No sir-e-e-e!" Juan shook his head and beat his chest, pretending to be Mr. Malcolm. "Mr. Malcolm told the miners:

" 'Now, boys, in all my years as fire boss, there's nothing I can't handle. Just leave her to me.'

"Whew! The miners were glad that Mr. Malcolm was with them. He went on talking:

" 'This here sow's no different than any other woman. I noticed if you're good to yer woman, she'll be good to you. This sow can't be that much different.'

"So Mr. Malcolm went right up to the bear and gave her a hug. . . ." Juan paused again, only this time he was grinning from ear to ear.

José and I exclaimed in disbelief:

"S-u-u-n!"

"No, sir! N-a-a-h!"

Dad held back a laugh, grinning instead. Ramona was animated. Mom and Angela weren't interested, but they were polite and didn't say anything.

"Not only that," Juan continued, "he told the bear:

" 'You're sure pretty. You have a pretty nose. And you have such shiny, curvy claws. And you have such a soft, brown coat.'

"Right away, the sow seemed to like Mr. Malcolm because he said so many pretty things to her. He could tell she wasn't a mean bear. She was just hungry. So Mr. Malcolm sez:

" 'Tell ya' what? Let's make a deal. Every day we'll leave you some food—outside—for you to eat. Only promise—don't come in the mine anymore. What say? Deal? Shake on it?' Mr. Malcolm reached his hand out. . . . The sow grabbed his hand. They shook on it. Right away, after they shook hands, the sow turned around and walked out of the mine. The next morning she came by. Mr. Malcolm had all kinds of food waiting for her. She ate it and went away."

"Na-a-a-w!" José doubted Juan's veracity. "That's a whopper if I ever heard one."

Juan stuck to his guns, pretending to be telling the truth and nothing but the truth. He didn't stammer nor let his voice quiver when he defended the story. He capped the story by quoting Mr. Malcolm:

" 'A nod and a handshake,' Mr. Malcolm said, 'is all a man needs in Chicorico.' "

"What a classy story!" I really liked Juan's story. "Someday, I'm going to be a miner. Maybe I can be a fire boss and have adventures like Mr. Malcolm."

"Hito, Mr. Malcolm told Juan a *chiste*," Dad explained. "Sometimes Mr. Malcolm likes to joke. He's good at pulling your leg. But, he's a very good fire boss, believe me. He's respected by the men."

Juan got defensive. He thought Dad accused him of lying. "But Dad, you're a fire boss too. Things like that *can* happen?"

"Oh, yes! He might've seen a bear. When he was fire bossing, he might've spotted a bear sniffing around the entry. When the bear didn't smell much—maybe some gas—she probably went along, away from the mine, searching for chokecherries, or berries she could eat."

"Aw, heck, Dad." Juan got recklessly brazen, "That's a boring story!"

Silence . . . stone-cold silence . . . a shock wave of silent tension rippled across the kitchen table. Juan's curt, brazen comment shocked us all. He actually insulted Dad. But we relaxed, realizing the insult had rolled right off Dad, like water rolling off a duck—lucky for Juan. Dad was more concerned about our romantic illusions of coal mining. He admonished us:

"You boys don't know what you're talking about. You think mining's glamorous!"

"But, Dad," I whined, "I want to be a miner like you. You have a swell job."

"¡A, mi hitos!" Mom surprised us. She was dismayed and just as opposed to mining as Dad. "You boys are too smart to be miners. We want you to get a good education, make something of yourselves."

"But Dad's smart," Juan defended Dad. "And he's a miner." He knew he had insulted Dad. Now, he was trying to vindicate himself by twisting Mom's words around.

Juan might be sly, but Dad was much smarter and wasn't easily fooled by him:

"Your mother's not saying I'm not smart. It's just you get old, your body wears out. It's better to work with your head. When I was probate judge, I made more money in two days than I did in a week in the mine. And I didn't work as hard. Here, let me show you my record." Dad turned to Angela, "Bring my book for 1940. It's on the first shelf of the library."

Dad was in politics for a while. The miners elected him to the state legislature for a two-year term, 1936–1938. Then he was elected probate judge for Colfax County for a two-year term. Angela brought the diary to Dad.

"*Mira, aquí*," Dad pointed at a column of figures. "For the month of May 1940:

Made money in the mine	$42.71
Made money in court	71.78
Marriages & Notary Public	23.80

"Now, look here, at July and August, the same thing. I worked harder in the mine and made less money. Money won't buy you happiness, but why kill yourself in the mine when you can have a better job with an education?"

The figures didn't lie. They jumped off the page, showing a real disparity. Working with your mind paid almost twice as much as working with your back. And you could work longer with your mind. Although working indoors all day didn't sound like much fun, either. Yet we pondered the figures as Dad read them aloud. When he reached the last column, Dad surprised us again. He actually married people.

"I didn't know you married people, Dad. You're not a priest." José studied the figures, especially the ones showing that Dad had performed marriages.

As a probate judge, Dad was authorized to probate wills, notarize papers, and conduct marriages. He translated the marriage ceremony into Spanish. He had Mrs. Chicarelli translate the ceremony into Italian, and she taught him how to read it aloud. For Slavic or Greek, he used English, although the Greeks often brought their own priest from Walsenburg to marry them. Most couples came from nearby Colorado or Texas. Both states required a blood test that took three days to process. New Mexico did not require the test. Instead of waiting three days to get married, couples from southern Colorado and west Texas came to Raton, where Dad married them. Nestled in the foothills of the Rocky Mountains, Raton also made an ideal honeymoon site.

"But? Only priests marry people." José was still confused about the difference between civil and religious marriages.

"I told everyone I married to go to the priest or a minister. I could only marry people under the law. I told them they weren't married in the eyes of God."

"*No importa*," Mom injected. "What's important is you kids need a good education to get somewhere—to make something of yourself. Already, Angela is talking of college to be a teacher. That's why we buy books for you to read. Have you read the new book? It came in the mail last week." Dad and Mom took every opportunity to buy books for us. Some were discarded from the Raton Public Library. Most were ordered from publishers who advertised in the union magazine, the *United Mine Workers of America Journal*.

Before we could answer Mom's question, Dad announced, "Tomorrow, Clara, after mass, I'll take the boys to the mine when no one's working."

"No! Manuelo! The mine's dangerous."

"The boys need to see the mine; mining's no work for a man. They'll see it's not so glamorous. . . . I'll take my fire boss lamp."

Mom wasn't placated. Yet our romantic illusions about coal mining worried her, although she didn't have anything against hard work. Mom didn't want us to grow up to be coal miners. She was too familiar with men whose bodies were wracked by the pain of arthritis and black lung disease from working too long in the cold, wet, and gassy coal mine. She was also too familiar with widowed mothers and children who had to fend for themselves, their husbands and fathers killed in the mine. Mom relented:

"Well, okay. I don't like it, Manuelo. Maybe it's best the boys see for themselves. Promise to be careful."

"I'll be careful," Dad promised as he lovingly gazed toward Mom.

Before Mom could change her mind, we all rose from the table, gave her a hug, and hopped to our after-supper chores straight away. Juan took warm water out of the boiler tank of the Majestic fogón. He washed and José dried the dishes. To replace the water they were using, I went to the pump outside and brought in two full buckets. I stood on a powder box and poured them into the boiler tank. We'd use it tomorrow. When done, I went into the living room.

Marilou Awiakta, "Baring the Atom's Mother Heart," from *Selu: Seeking the Corn-Mother's Wisdom* (1993)

b. 1936

Having grown up in the shadow of Tennessee's Oak Ridge—established during the 1940s as a base for the Manhattan Project, the secretive government operation that developed the atomic bombs eventually dropped on Hiroshima and Nagasaki—Cherokee poet Marilou Awiakta's writing is inflected by her lifelong relationship to, and ambivalence about, nuclear energy. Awiakta's unique fusion of her Cherokee and Appalachian heritage with science has brought her international recognition. In 1985 the U.S. Information Agency chose her books *Abiding Appalachia: Where Mountain and Atom Meet* (1978) and *Rising Fawn and the Fire Mystery* (1983) for the global tour of their exhibit “Women in the Contemporary World.” Awiakta’s third book, *Selu: Seeking the Corn-Mother’s Wisdom*, applies Native American philosophy to contemporary environmental issues. A quote from *Selu* is engraved in the River Wall of the Bicentennial Capitol Mall in Nashville, and Awiakta’s poem “Motheroot” is lined in marble along one border of the new Fine Arts Walkway at the University of California, Riverside.

Discussion Generators

In atomic technology Awiakta finds a metaphor for the interconnectedness and balance of all living things on Earth. “Nuclear science,” she explains, “reveals how this balance operates in the invisible dimension, the heart of nature. On a subatomic level, the world is image and shadow, not solid mass. It’s constantly moving, full of light.” In the essay “Baring the Atom’s Mother Heart,” Awiakta defines the atom as the “nurturing energy of the universe” and warns of human efforts to manipulate such all-penetrating power. What does Awiakta mean when she characterizes the atom as possessing a “mother heart,” and in what way does she suggest that its power is “nurturing”? How does this feminine dimension of nuclear energy contrast with its destructive—and, Awiakta suggests, masculine—side? Do you find Awiakta’s gendered dualisms helpful in understanding the paradox of nuclear energy or are they problematic?

“What is the atom, Mother? Will it hurt us?”

I was nine years old. It was December 1945. Four months earlier, in the heat of an August morning—Hiroshima. Destruction. Death. Power beyond belief, released from something invisible. Without knowing its name, I’d already felt the atom’s power in another form. Since 1943, my father had commuted eighteen miles from our apartment in Knoxville to the plant in Oak Ridge—the atomic frontier where the atom had been split, where it still was splitting. He left before dawn and came home long after dark. “What do you do, Daddy?”—“I can’t tell you, Marilou. It’s part of something for the war. I don’t know what they’re making out there or how my job fits into it.”

“What’s inside the maze?”

“Something important . . . and strange. I see long, heavy trucks coming in. What they’re bringing just seems to disappear. Somebody must know what happens to it, but nobody ever talks about it. One thing for sure—the government doesn't spend millions of dollars for nothing. It's something big. I can't imagine what."

I couldn't either. But I could feel its energy like a great hum.

Then, suddenly, it had an image: the mushroom cloud. It had a name: the atom. And our family was then living in Oak Ridge. My father had given me the facts. I also needed an interpreter.

"*What is the atom, Mother? Will it hurt us?*"

"It can be used to hurt everybody, Marilou. It killed thousands of people in Hiroshima and Nagasaki. But the atom itself . . . ? It's invisible, the smallest bit of matter. And it's in everything. Your hand, my dress, the milk you're drinking—all of it is made with millions and millions of atoms and they're all moving. But what the atom means . . . ? I don't think anyone knows yet. We have to have reverence for its nature and learn to live in harmony with it. Remember the burning man."

"I remember." When I was six years old, his screams had brought my mother and me running to our front porch. Mother was eight months pregnant. What we saw made her hold me tight against her side. Across the street, in the small parking lot of the dry cleaner's, a man in flames ran, waving his arms. Another man chased him, carrying a garden hose turned on full force, and shouting, "Stop, stop!" The burning man stumbled and sank to his knees, shrieking, clawing the air, trying to climb out of his pain. When water hit his arms, flesh fell off in fiery chunks. As the flames went out, his cries ceased. He collapsed slowly into a charred and steaming heap.

Silence. Burned flesh. Water trickling into the gutter . . .

The memory flowed between Mother and me, and she said, as she had said that day, "Never tempt nature, Marilou. It's the nature of fire to burn. And of cleaning fluid to flame near heat. The man had been warned over and over not to work with the fluid, then stoke the furnace. But he kept doing it. Nothing happened. He thought he was in control. Then one day a spark . . . The atom is like the fire."

"So it *will* hurt us."

"That depends on us, Marilou."

I understood. Mother already had taught me that beyond surface differences, everything is in physical and spiritual connection—God, nature, humanity. All are one, a circle. It seemed natural for the atom to be part of this connection. At school, when I was introduced to Einstein's theory of relativity—that energy and matter are one—I accepted the concept easily.

Peacetime brought relaxation of some restrictions in Oak Ridge. I learned that my father was an accountant. The "long, heavy trucks" brought uranium ore to the graphite reactor, which was still guarded by a maze of fences. The reactor reduced the ore to a small amount of radioactive material. Safety required care and caution. Scientists called the reactor "The lady" and, in moments of high emotion, referred to her as "our beloved reactor."

"What does she look like, Daddy?"

"They tell me she has a seven-foot shield of concrete around a graphite core, where the atom is split." I asked the color of graphite. "Black," he said. And I imagined a great, black queen, standing behind her shield, holding the splitting atom in the shelter of her arms.

I also saw the immense nurturing potential of the atom. There was intensive research into fuels, fertilizers, mechanical and interpretative tools. Crops and animals were studied for the effects of radiation. Terminal cancer patients came from everywhere to the research hospital. I especially remember one newspaper picture of a man with incredibly thin hands reaching for the "atomic cocktail" (a container of radioactive isotopes). His face was lighted with hope.

At school we had disaster drills in case of nuclear attack (or in case someone got careless around the reactor). Scientists explained the effects of an explosion from "death light" to fallout. They also emphasized the peaceful potential of the atom and the importance of personal commitment in using it. Essentially, their message was the same as my mother's. "If we treat the atom with reverence, all will be well."

But all is not well now with the atom. The arms race, the entry of Big Business into the nuclear industry, and accidents like Three Mile Island cause alarm. Along with me, women protest, organize anti-nuclear groups, speak out. But we must also take time to ponder woman's affinities with the atom and to consider that our responsibilities for its use are more profound than we may have imagined.

We should begin with the atom itself, which is approximately two trillion times smaller than the point of a pin. We will focus on the nature and movement of the atom, not on the intricacies of nuclear physics. To understand the atom, we must flow with its pattern, which is circular.

During the nineteenth and twentieth centuries, scientists theorized about the atom, isolated it, discovered the nucleus, with its neutrons, protons, electrons. The atom appeared to resemble a Chinese nesting ball—a particle within a particle. Scientists believed the descending order would lead to the ultimate particle—the final, tiny bead. Man would penetrate the secret of matter and dominate it. All life could then be controlled, like a machine.

Around the turn of the century, however, a few scientists began to observe the atom asserting its nature, which was more flexible and unpredictable than had been thought. To explain it required a new logic, and, in 1905, Einstein published his theory of relativity. To describe the atom also required new use of language in science because our senses cannot experience the nuclear world except by analogy. The great Danish physicist, Niels Bohr, said, "When it comes to atoms, language can be used only as in poetry. The poet, too, is not nearly so concerned with describing facts as with creating images and mental connections."

As research progressed, the word *mystery* began to appear in scientific writing, along with theories that matter might not end in a particle after all. Perhaps the universe resembled a great thought more than a great machine. The linear path was bending . . . and in the mid-1970s the path ended in an infinitesimal circle: the quark. A particle so small that even with the help of huge machines, humans can see only its trace, as we see the vapor trail of an airplane in the stratosphere. A particle ten to one hundred million times smaller than the atom. Within the quark, scientists now perceive matter refining beyond space-time into a kind of mathematical operation, as nebulous and real as an unspoken thought. It is a mystery that no conceivable research is likely to dispel, the life force in process—nurturing, enabling, enduring, fierce.

I call it the atom's mother heart.

Nuclear energy is the nurturing energy of the universe. Except for stellar explosions, this energy works not by fission (splitting) but by fusion—attraction and melding. With the relational process, the atom creates and transforms life. Women are part of this life force. One of our natural and chosen purposes is to create and sustain life—biological, mental and spiritual.

Women nurture and enable. Our "process" is to perceive relationships among elements, draw their energies to the center and fuse them into a whole. Thought is our essence; it is intrinsic for us, not an aberration of our nature, as Western tradition often asserts.

Another commonality with the atom's mother heart is ferocity. When the atom is split—when her whole is disturbed—a chain reaction begins that will end in an explosion unless the reaction is contained, usually by a nuclear reactor. To be productive and safe, the atom must be restored to its harmonic, natural pattern. It has to be treated with respect. Similarly, to split woman from her thought, sexuality and spirit is unnatural. Explosions are inevitable unless wholeness is restored.

In theory, nature has been linked to woman for centuries—from the cosmic principle of the Great Mother-Goddess to the familiar metaphors of Mother Nature and Mother Earth. But to connect the life force with *living* woman is something only some ancient or so-called "primitive" cultures have been wise enough to do. The linear, Western, masculine mode of thought has been too intent on conquering nature to learn from her a basic truth: *To separate the gender that bears life from the power to sustain it is as destructive as to tempt nature herself.*

This obvious truth is ignored because to accept it would acknowledge woman's power, upset the concept of woman as sentimental—passive, all-giving, all-suffering—and disturb public and private patterns. But the atom's mother heart makes it impossible to ignore this truth any longer. She is the interpreter of new images and mental connections not only for humanity, but most particularly for women, who have profound responsibilities in solving the nuclear dilemma. We can do much to restore harmony. But time is running out. . . .

Shortly after Hiroshima Albert Einstein said, "The unleashed power of the atom has changed everything save our modes of thought, and thus we drift toward unparalleled catastrophe." Now, deployment of nuclear missiles is increasing. A going phrase in Washington is, "When the war starts . . ." Many nuclear power plants are being built and operated with money, not safety, as the bottom line. In spite of repeated warnings from scientists and protests from the public, the linear-thinking people continue to ignore the nature of the atom. They act irreverently. They think they're in control. One day a spark. . . .

I look beyond the spectres of the burning man and the mushroom cloud to a time two hundred years ago, when destruction was bearing down on the Cherokee nation. My foremothers took their places in the circles of power along with the men. Outnumbered and outgunned, the nation could not be saved. But the Cherokee and their culture survived—and women played a strong part in that survival.

Although the American culture is making only slow progress toward empowering women, there is much we can do to restore productive harmony with the atom. Protest and litigation are important in stopping nuclear abuse, but total polarization between pro- and anti-nuclear people is simplistic and dangerous. It is not true that all who believe in nuclear energy are bent on destruction. Neither is it true that all who oppose it are "kooks" or "against progress." Such linear, polar thinking generates so much anger on both sides that there is no consensual climate where reasonable solutions can be found. The center cannot hold. And the beast of catastrophe slouches toward us. We need a network of the committed to ward it off. Women at large can use our traditional intercessory skills to create this network through organizations, through education and through weaving together conscientious protagonists in industry, science and government. Women who are professionals in these fields should share equally in policy making.

Our energies may fuse with energies of others in ways we cannot foresee. I think of two groups of protesters who came to Diablo Canyon, California, in the fall of 1981. Women and men protested the activation of a nuclear power plant so near an earthquake fault. The first group numbered nearly three thousand. The protest was effective, but it says much about the dominant, holistic mode of American thought that an article about the second group was buried in the middle of a San Francisco newspaper.

After the three thousand had left Diablo Canyon to wind and silence, a band of about eighty Chumash Indians came to the site of the power plant. They raised a wood-sculptured totem and sat in a circle around it for a daylong prayer vigil. Jonathan Swift Turtle, a Mewok medicine man, said that the Indians did not oppose nuclear technology but objected to the plant's being built atop a sacred Chumash burial site as well as near an earthquake fault. He said he hoped the vigil would bring about "a moment of harmony between the pro- and anti-nuclear factions."

The Chumash understand that to split the atom from the sacred is a deadly fission that will ultimately destroy nature and humanity. I join this circle of belief with an emblem I created for my life and work—the sacred white deer of the Cherokee leaping in the heart of the atom. My ancestors believed that if a hunter took the life of a deer without asking its spirit for pardon, the immortal Little Deer would track the hunter to his home and cripple him. The reverent hunter evoked the white deer's blessing and guidance.

For me, Little Deer is a symbol of reverence. Of hope. Of belief that if we humans relent our anger and create a listening space, we may attain harmony with the atom in time. If we do not, our world will become a charred and steaming heap. Burned flesh. Silence . . .

There will be no sign of hope except deep in the invisible, where the atom's mother heart—slowly and patiently—bears new life.

Wendy Rose, "Plutonium Vespers" (1994)

b. 1948

Poet, nonfiction writer, artist, educator, and anthropologist Wendy Rose was born Bronwen Elizabeth Edwards to parents of mixed racial identity. Though she is partly of Hopi and Miwok ancestry, she was raised in a predominately white community in California's San Francisco Bay Area. Growing up in an urban environment far removed from reservation life and Native American relatives provided her little access to her native roots as a child. Rose's poetry, collected in volumes such as *Bone Dance* (1994), *Going to War With All My Relations* (1993), and *Lost Copper* (1980), often reflects the challenges she has faced reconciling her multiple ethnic identities. In her writings, Rose often challenges the appropriation of Native American culture, including "whiteshamanism," a term she applies to white writers pretending to be so entrenched in Native American ways and beliefs that they are "just as Indian" as those born into the culture.

Discussion Generators

The poem printed here, "Plutonium Vespers," addressed to her grandparents, articulates Rose's sense of kinship with her native ancestors and the natural world she associates with them, while at the same time expressing her outrage toward her white ancestors whose actions—including the dropping of the atomic bomb—have resulted in ecological destruction. Aside from the title, what words and phrases suggest that this poem has a "nuclear" theme? What do you suppose the speaker's purpose might be in addressing the poem to ancestors and ending the poem by invoking them a second time? What is the effect of the repetition of "pity" in the penultimate stanza? Does it evoke pity in the reader or does it elicit some other response?

Grand mother
Grand father

take this offering
of flesh, this color
and this color, take
all the memories,
take the pain,
take it and shake it
everywhere shake it
all of us shaking
I am shaking

I hear you in a whisper
cricket's voice and thunder;
I hear you warble, rasp,
be in the bullroarer
mimicking hummingbird's wings.
I hear you in woodpecker's laugh,
the creaking wood floor,
silent swift dive of bat;

I hear you in the bending reeds,
brittle oak leaves, the furry moss.

Pity these ones, pity them,
pity these ones, pity the skin
that hangs in ribbons, pity
the round eyes, melted tongue,
pity the sunlight stolen,
white grass uncoiling underground;
pity these ones
who have been touched
by untouchable stuff.
Pity them in the fearsome rain,
pity them who shit great bombs onto earth,
pity the woman who stepped
on the serpent,
pity the quivering sky, the stars
of the milky way, creation
as it unravels, pity us shaking

this small offering is for you,
this small thing, black
willow stick, smoldering sage,
granite crumbling into sand,
the wayward bug, the scorched bone,
for you
Grandmother
Grandfather

Chapter Four

Alternatives: Wind, Water, Earth, Sun

David Lee, "Remnant" (2008)

b. 1944

In 1997, David Lee was named the first Poet Laureate of the state of Utah, celebrating a career encompassing three decades of teaching literature and writing at Southern Utah University, as well as the publication of fifteen books of poetry. His work draws on his rich and diverse life experiences that include seminary study, boxing, raising hogs, playing semi-pro baseball, and earning a Ph.D. with a concentration in the poetry of John Milton. Of his own poetry, Lee has said, "Art does not imitate life. Life imitates art. Art sets the model, and if we accept the model, then we change our lives to fit that. What I'm trying to do is create a blueprint for a pattern of dignity."

Discussion Generators

"Remnant," a short poem from Lee's 2010 collection *Stone Wind Water*, offers insight into the poet's scheme for seeing nature. In Lee's vision, a "Michelangelo effect" exists between ourselves and the natural world, in which various elements—wind, stone, animal, poet—evoke the best in each other, helping them to come closer to their ideal selves. How can this poetic notion be a "blueprint for a pattern of dignity," as Lee proposes? Contrast this free-form vision of wind with Susan McLean's poem in this chapter. How might a poem about wind help us think more deeply about wind as a source of energy?

being the remainder of a carved down pretentious sonnet on
the Michelangelo effect of wind, including a joyous horse and a

gorgeous arch, both unanticipated before being stumbled upon
while wandering

A wind gust
          wallows
in an arroyo,
          rolls,
then rises and shakes its mane,
leaps
      to trample a rabbit brush
huddling on the lip.

Suddenly
this crimson flaming arch,
wind's triumphant joy,
the majesty of sandstone
frozen momentarily,
a grand pas de deux,
in its slow sojourn
back
to bright dust.

Cate Marvin, "A Windmill Makes a Statement" (2001)

b. 1969

The only child of a C.I.A. intelligence analyst and an editor for the Crime Prevention Council, Cate Marvin grew up in Washington, D.C., and graduated from Vermont's Marlboro College. She earned an MFA in poetry from the University of Houston, an MFA in fiction from the Iowa Writers' Workshop, and a Ph.D. in English and comparative literature from the University of Cincinnati. Marvin's first book, *World's Tallest Disaster*, was chosen by then–U.S. Poet Laureate Robert Pinksy for the 2000 Kathryn A. Morton Prize. Her poems have appeared in *Tin House, Virginia Quarterly Review, The New England Review*, *Poetry*, *The Kenyon Review*, and many other journals and magazines. A *Publishers Weekly* review of Marvin's work referred to her as a "postmodern Plath," noting, "Marvin can make you laugh at crying and cry at laughing . . . few works so rife with satire ever took the human condition more seriously."

Discussion Generators

Marvin's title for the poem printed below declares that the poem's speaker, a windmill, "makes a statement." What sort of statement is this windmill making? What is the content of its message? Its emotional tone? How might the statement shift if the speaker were some other source of energy—a nuclear plant, for instance? A hydroelectric dam? A solar panel? Try your hand at composing your own original poem from the perspective of an energy source of your choice. What did you learn from this exercise?

You think I like to stand all day, all night,
all any kind of light, to be subject only
to wind? You are right. If seasons undo
me, you are my season. And you are the light
making off with its reflection as my stainless
steel fins spin.

On lawns, on lawns we stand,
we windmills make a statement. We turn air,
churn air, turning always on waiting for your
season. There is no lover more lover than the air.
You care, you care as you twist my arms
round, till my songs become popsicle

and I wing out radiants of light all across
suburban lawns. You are right, the churning
is for you, for you are right, no one but you
I spin for all night, all day, restless for your

sight to pass across the lawn, tease grasses,
because I so like how you lay above me,

how I hovered beneath you, and we learned
some other way to say: *There you are.*

You strip the cut, splice it to strips, you mill
the wind, you scissor the air into ecstasy until
all lawns shimmer with your bluest energy.

Susan McLean, "Reaping the Wind" (2010)

b. 1953

Susan McLean has published in such journals as *Kalliope* and *The Classical Outlook* and won the Richard Wilbur Award as well as a McKnight Fellowship. McLean is a professor of English at Southwest Minnesota State University in Marshall, located near the Buffalo Ridge Wind Towers, erected in 1993. McLean describes her experience living in close proximity to one of the largest wind farms in the United States: "When I first saw the Buffalo Ridge Wind Farm about thirty miles from my home, I was struck by the giant size of the wind turbines, which towered above the old-fashioned windmills on farms below them and made the hills nearby look like a vision out of science fiction. At first I was pleased by the thought of clean energy being generated nearby, but also glad that the turbines weren't closer, since the constant motion of the blades was unsettling to watch. I have since learned from my students' research papers the hidden costs of wind energy, such as the deaths of birds and bats who fly too near the towers, while the turbines themselves have proliferated so much that I can see them without even leaving town."

Discussion Generators

"Reaping the Wind" suggests a tension between aesthetics and alternative energy that may surprise some who find all aspects of wind energy attractive. How do the formal structure and appeal to ritual that McLean employs match or contrast with the movements that turbines make? McLean's venue of publication is also an interesting choice in the context of energy policy and literature. *First Things* has a stated mission of "religiously informed public philosophy for the ordering of society." What kinds of societal and spiritual assertions do you see at work in this compressed, powerful poem?

The giants swing their triple arms or poise,
frozen like hazmat signs, on every hill:
alien prayer wheels of unending noise
or monuments to birds and bats they kill;
Shivas, whose dance preserves us and destroys
a landscape that no longer can stand still.

George Eliot, "Outside Dorlcote Mill," from *The Mill on the Floss* (1860)

1819–1880

Mary Anne Evans, commonly known by her pen name George Eliot, was an English novelist, journalist, and translator and was one of the most important writers of the Victorian era. She is the author of seven novels, including *The Mill on the Floss* (1860), *Silas Marner* (1861), *Middlemarch* (1872), and *Daniel Deronda* (1876), most of them set in rural England and celebrated for their realism and psychological insight. Eliot often based the settings of her novels on places from her childhood; the Dorlcote Mill in *The Mill on the Floss*, for instance, was based upon Griff House, now a hotel, in the coal-mining region of her native Warwickshire.

Discussion Generators

The following excerpt from *The Mill on the Floss* shows the centrality of water-powered industrial mills in nineteenth-century English life. What sensory details does Eliot use to depict the Dorlcote Mill, and what emotional effect do these have upon the reader? Is this a peaceful scene? A noisy scene? Something else? Does Eliot's description make water power seem an appealing alternative to other ways of producing energy?

A wide plain, where the broadening Floss hurries on between its green banks to the sea, and the loving tide, rushing to meet it, checks its passage with an impetuous embrace. On this mighty tide the black ships—laden with the fresh-scented fir-planks, with rounded sacks of oil-bearing seed, or with the dark glitter of coal—are borne along to the town of St Ogg's, which shows its aged, fluted red roofs and the broad gables of its wharves between the low wooded hill and the river brink, tinging the water with a soft purple hue under the transient glance of this February sun. Far away on each hand stretch the rich pastures, and the patches of dark earth, made ready for the seed of broad-leaved green crops, or touched already with the tint of the tender-bladed autumn-sown corn. There is a remnant still of the last year's golden clusters of beehive ricks rising at intervals beyond the hedgerows; and everywhere the hedgerows are studded with trees: the distant ships seem to be lifting their masts and stretching their red-brown sails close among the branches of the spreading ash. Just by the red-roofed town the tributary Ripple flows with a lively current into the Floss. How lovely the little river is, with its dark, changing wavelets! It seems to me like a living companion while I wander along the bank and listen to its low placid voice, as to the voice of one who is deaf and loving. I remember those large dipping willows. I remember the stone bridge.

And this is Dorlcote Mill. I must stand a minute or two here on the bridge and look at it, though the clouds are threatening, and it is far on in the afternoon. Even in this leafless time of departing February it is pleasant to look at—perhaps the chill damp season adds a charm to the trimly kept, comfortable dwelling-house, as old as the elms and chestnuts that shelter it from the northern blast. The stream is brimful now, and lies high in this little withy plantation, and half drowns the grassy fringe of the croft in front of the house. As I look at the full stream, the vivid grass, the delicate bright-green powder softening the outline of the great trunks and branches that gleam from under the bare purple boughs, I am in love with moistness, and envy the white ducks that are dipping their heads far into the water here among the withes, unmindful of the awkward appearance they make in the drier world above.

The rush of the water, and the booming of the mill, bring a dreamy deafness, which seems to heighten the peacefulness of the scene. They are like a great curtain of sound, shutting one out from the world beyond. And now there is the thunder of the huge covered waggon coming home with sacks of grain. That honest waggoner is thinking of his dinner, getting sadly dry in the oven at this late hour; but he will not touch it till he has fed his horses,—the strong, submissive, meek-eyed beasts, who, I fancy, are looking mild reproach at him from between their blinkers, that he should crack his whip at them in that awful manner, as if they needed that hint! See how they stretch their shoulders up the slope towards the bridge, with all the more energy because they are so near home. Look at their grand shaggy feet that seem to grasp the firm earth, at the patient strength of their necks, bowed under the heavy collar, at the mighty muscles of their struggling haunches! I should like well to hear them neigh over their hardly-earned feed of corn, and see them, with their moist necks freed from the harness, dipping their eager nostrils into the muddy pond. Now they are on the bridge, and down they go again at a swifter pace, and the arch of the covered waggon disappears at the turning behind the trees.

Now I can turn my eyes towards the mill again, and watch the unresting wheel sending out its diamond jets of water. That little girl is watching it too: she has been standing on just the same spot at the edge of the water ever since I paused on the bridge. And that queer white cur with the brown ear seems to be leaping and barking in ineffectual remonstrance with the wheel; perhaps he is jealous, because his playfellow in the beaver bonnet is so rapt in its movement. It is time the little playfellow went in, I think; and there is a very bright fire to tempt her: the red light shines out under the deepening grey of the sky. It is time, too, for me to leave off resting my arms on the cold stone of this bridge. . . .

Ah, my arms are really benumbed. I have been pressing my elbows on the arms of my chair, and dreaming that I was standing on the bridge in front of Dorlcote Mill, as it looked one February afternoon many years ago. Before I dozed off, I was going to tell you what Mr and Mrs Tulliver were talking about, as they sat by the bright fire in the left-hand parlour, on that very afternoon I have been dreaming of.

Joan Didion, "At the Dam" (1970)

b. 1934

Best known for her novels and her literary journalism, Joan Didion received the National Book Award for Nonfiction in 2005 for *The Year of Magical Thinking*. Didion's novels and essays explore cultural entropy and the disintegration of American morals, in which the overriding theme is individual and social fragmentation. In 2007, the National Book Foundation awarded

Didion its annual Medal for Distinguished Contribution to American Letters, the citation for which reads: "An incisive observer of American politics and culture for more than forty-five years, her distinctive blend of spare, elegant prose and fierce intelligence has earned her books a place in the canon of American literature as well as the admiration of generations of writers and journalists." The following essay, "At the Dam," explores the ways that engineering projects are representative of the America at the heart of Didion's writing—a country of both vast potential and dizzying hubris.

Discussion Generators

Didion's short essay recounts the author's visit to Hoover Dam, one of the world's largest hydroelectric power stations. In the essay Didion suggests that at the "innocent" time of its construction, Hoover Dam was emblematic of the notion that "mankind's brightest promise lay in American engineering." In what ways does Didion's essay reinforce this sense of optimism about the potential of large-scale hydroelectric energy projects? In what ways does the essay seem to challenge this trust in engineering to solve humanity's problems? What do you think Didion means when she refers to "something beyond energy, beyond history" at Hoover Dam? What kind of force, or energy, is she attempting to describe?

Since the afternoon in 1967 when I first saw Hoover Dam, its image has never been entirely absent from my inner eye. I will be talking to someone in Los Angeles, say, or New York, and suddenly the dam will materialize, its pristine concave face gleaming white against the harsh rusts and taupes and mauves of that rock canyon hundreds or thousands of miles from where I am. I will be driving down Sunset Boulevard, or about to enter a freeway, and abruptly those power transmission towers will appear before me, canted vertiginously over the tailrace. Sometimes I am confronted by the intakes and sometimes by the shadow of the heavy cable that spans the canyon and sometimes by the ominous outlets to unused spillways, black in the lunar clarity of the desert light. Quite often I hear the turbines. Frequently I wonder what is happening at the dam this instant, at this precise intersection of time and space, how much water is being released to fill downstream orders and what lights are flashing and which generators are in full use and which just spinning free.

I used to wonder what it was about the dam that made me think of it at times and in places where I once thought of the Mindanao Trench, or of the stars wheeling in their courses, or of the words *As it was in the beginning, is now and ever shall be, world without end, amen.* Dams, after all, are commonplace: we have all seen one. This particular dam had existed as an idea in the world's mind for almost forty years before I saw it. Hoover Dam, showpiece of the Boulder Canyon project, the several million tons of concrete that made the Southwest plausible, the *fait accompli* that was to convey, in the innocent time of its construction, the notion that mankind's brightest promise lay in American engineering.

Of course the dam derives some of its emotional effect from precisely that aspect, that sense of being a monument to a faith misplaced. "They died to make the desert bloom," reads a plaque dedicated to the 96 men who died building this first of the great high dams, and in this context the worn phrase touches, suggests all of that trust in harnessing resources, in the meliorative power of the dynamo, so central to the early Thirties. Boulder City, built in 1931 as the construction town for the dam, retains the ambience of a model city, a new town, a toy triangular grid of green lawns and trim bungalows, all fanning out from the Reclamation building. The bronze sculptures at the dam itself evoke muscular citizens of a tomorrow that never came, sheaves of wheat clutched heavenward, thunderbolts defied. Winged Victories guard the flagpole. The flag whips in the canyon wind. An empty Pepsi-Cola can clatters across the terrazzo. The place is perfectly frozen in time.

But history does not explain it all, does not entirely suggest what makes the dam so affecting. Nor, even, does energy, the massive involvement with power and pressure and the transparent sexual overtones to that involvement. Once when I revisited the dam I walked through it with a man from the Bureau of Reclamation. For a while we trailed behind a guided tour, and then we went on, went into parts of the dam where visitors do not generally go. Once in a while he would explain something, usually in that recondite language having to do with "peaking power," with "outages" and "dewatering," but on the whole we spent the afternoon in a world so alien, so complete and so beautiful unto itself that it was scarcely necessary to speak at all. We saw almost no one. Cranes moved above us as if under their own volition. Generators roared. Transformers hummed. The gratings on which we stood vibrated. We watched a hundred-ton steel shaft plunging down to that place where the water was. And finally we got down to that place where the water was, where the water sucked out of Lake Mead roared through thirty-foot penstocks and then into thirteen-foot penstocks and finally into the turbines themselves. "Touch it," the Reclamation man said, and I did, and for a long time I just stood there with my hands on the turbine. It was a peculiar moment, but so explicit as to suggest nothing beyond itself.

There was something beyond all that, something beyond energy, beyond history, something I could not fix in my mind. When I came up from the dam that day the wind was blowing harder, through the canyon and all across the Mojave. Later, toward Henderson and Las Vegas, there would be dust blowing, blowing past the Country-Western Casino FRI & SAT NITES and blowing past the Shrine of Our Lady of Safe Journey STOP & PRAY, but out at the dam there was no dust, only the rock and the dam and a little greasewood and a few garbage cans, their tops chained, banging against a fence. I walked across the marble star map that traces a sidereal revolution of the equinox and fixes forever, the Reclamation man had told me, for all time and for all people who can read the stars, the date the dam was dedicated. The star map was, he had said, for when we were all gone and the dam was left. I had not thought much of it when he had said it, but I thought of it then, with the wind whining and the sun dropping behind a mesa with the finality of a sunset in space. Of course that was the image I had seen always, seen it without quite realizing what I saw, a dynamo finally free of man, splendid at last in its absolute isolation, transmitting power and releasing water to a world where no one is.

John Calderazzo, "Into the Ring of Fire" (2004)

b. 1946

In order to get a sense of what everyday life is like in the vicinity of a volcano—a presence at once magnificent and threatening—essayist and nature writer John Calderazzo traveled to a number of the world's volcanic regions for his 2004 book, *Rising Fire: Volcanoes and Our Inner Lives*, in which the essay below was published. Such places include Italy's Mount Etna; Parícutin, Mexico, where in 1943 a volcano suddenly erupted in a cornfield; the small volcano-dominated Caribbean island of Montserrat; California's Mount Shasta, a magnet for the mystically inclined; and Hawaii's Kilauea, discussed in the following piece. His observations of the ways volcanoes have shaped humankind's sense of the sacred and his profiles of various volcano fanatics are as compelling as his dramatic accounts of volcanic eruptions. The practice of human sacrifice to appease the gods of the underworld, scientific breakthroughs, and the cosmic sense of awe and impermanence that volcanoes inspire—all of these are addressed in Calderazzo's writing.

Discussion Generators

In the selection from *Rising Fire*, we learn about several people who are obsessed with volcanoes, including Maurice Krafft, who calls himself "sort of a crazy witch doctor" only partly tongue-in-cheek. Can you understand their motivations to get close to such dangerous and wild places, or are they inscrutable on some level? Opposed to the sun, the energy from volcanoes comes from underfoot. Is there a critical difference here or not? How might this meditation on volcanic energy contribute to our understanding of geothermal power?

Nobody had seen more of that wild uncertainty, at unbelievably close range, than the pair of ghosts I'd been tracking. I don't mean Pele and her consort Kamapua'a. The ghosts I was after had also been married, but much more happily, since their volcanic passions complemented each other rather than clashed. Their names were Maurice and Katia Krafft, ordinary mortals but the world's most famous—or notorious—volcano watchers and photographers.

Ever since I'd started to hear stories of them a few years earlier, I'd been intrigued by their passion for volcanoes, by their desire and willingness to hike to the edge of a blowing crater while shattered, glowing rocks thudded down around them. Although their behavior sounded as if it bordered on insanity, they weren't nut cases at all, volcanologists I'd talked to had assured me. But they were scientists who really believed in *hands-on experience.*

And they loved the Big Island. Entirely by coincidence, I had booked SueEllen and myself into a lovely old country house called My Volcano, a B&B in Volcano village that turned out to be where the Kraffts had frequently stayed—one unit had been wired especially for their fax machine and other equipment. One morning, over a breakfast of pancakes and macadamia nuts, I leafed through a few of their signed photography books. The B&B owner, white-haired and rugged-looking Gordon Morse, was a lifetime Hawaiian and volcano fan himself. Once, about thirty years before, while eating a cucumber sandwich one afternoon on a slope of Kilauea, he had heard a rumble and seen great red sheets of lava spurt without warning from the earth. He understood what the Kraffts were about. Now that I had stood on the lip of a skylight and seen lava rivering right under my feet, I think I did, too. I was beginning to understand that edgy state of mind in which deep interest or fascination can pitch over into obsession.

Apparently, in Italy, I had been following in the footsteps of the Kraffts for some time without realizing it. After I had flown home from Rome, I watched a video in which Maurice and Katia took turns talking to the camera. Their marriage was fate, they implied, and the fate first struck at Stromboli volcano, years before they knew each other. In the 1960s, they had met at the University of Strasbourg, on the French-German border, where Maurice was studying geology and Katia geochemistry. But quite independently, each had fallen in love with the third party in their relationship—erupting volcanoes—while on childhood family trips to Stromboli. They had seen the lava fly in the night, and became instantly inspired by "mountains *alive*," as Maurice liked to call volcanoes, and that was that.

As graduate students, they taught themselves to shoot pictures and videos, mainly as a way to finance their more and more ambitious expeditions. Maurice, who was gifted at public relations, liked to explain that they had needed to eat, and after all, "to eat stone is quite difficult. We have tried it!" Often working for days without sleep while cinders bounced off their hard hats, they filmed and took field notes and measurements: at Etna, Réunion Island in the Indian Ocean, Mount Augustine in Alaska, and dozens of other eruption sites. Their skills as scientists and artists improved fast, and soon they were completely self-supporting. They had no children, no teaching or research obligations. After seventeen books, five exuberant films, and increasingly popular European lecture tours, they'd turned into Emmy winners and the world's most swashbuckling volcano photographers.

They knew almost everyone who crawled on or near simmering craters, or so it seemed, and the telephone and the fax machine in their Rhine Valley home took in messages day and night. "Hurry!" a colleague in some remote jungle might say. "It's going to blow!" Within hours, their gas masks, light meters, and cameras clanking around their necks, they would jet off to Mount Agung on Java, El Chichón in Mexico, or anywhere the heat of the restless earth broke out in photogenic fire. Toward the end of the 1980s, their major goal in life had crystallized: to document the great eruptions of our time. *All* of them. In 1988 alone, they circled the globe eight times.

In the words of their friend and fellow volcanologist Bob Tilling, of the United States Geological Survey, Maurice and Katia had become "the Jacques Cousteaus of volcanology." Or as a lot of their friends affectionately called them, "The Volcano Devils." But some people who saw them in action called them crazy. They were certainly eruption addicts, apocalypse junkies, with probably much more desire than true scientific need to creep as close as they could to rising fire, and they had climbed more than 200 active volcanoes, the places they called "earth giving birth."

You've probably seen their work on television and in magazines. They built a photo archive of 300,000 volcano images and compiled a wonderful collection of volcano art and rare books and films. By 1991, the last year of their lives, they had become scientists with the souls of artists, souls of fire. They were both still in their forties.

Like many geologists, they were also physically tough. Ninety-five-pound Katia sometimes strapped on a backpack that looked like it weighed as much as she did. In their graduate student days, she and Maurice would load their gas-measuring equipment and camping gear onto bicycles, and up through muddy jungle paths they'd go. Or they'd crunch along for miles over just-hardened, jumbled-up, black glass lava, then spend the night huddled in a pup tent in howling winds at eighteen or twenty thousand feet.

They'd trudge right up to a lava geyser spurting a quarter-mile high, uncap their lenses, and get to work. Clunking along in heat-resistant Tin Man suits they'd designed themselves, or wearing pointed aluminum helmets that sloped over their shoulders to deflect lava-spatter, they'd approach a crater gloved hand in gloved hand. Curly-haired, bearish Maurice would scan the sulfurous sky for whistling lava bombs while slender Katia studied the ground for orange cracks or rivulets of 2,000-degree Fahrenheit rock flows. Filmed by an assistant from a much safer distance, they appeared to be dancing before a curtain of flames.

Lava sometimes melted the soles of their boots. Heat warped their camera lenses. If they weren't careful, gases that smelled as bad or worse than the stuff I'd gotten a whiff of could "process" the film in their cameras long before they got it back to the lab. Their cameras grew so hot they sometimes had to wear tinfoil masks to keep their cheekbones from getting singed.

Once, in a crater in Java, they launched a rubber raft on a 600-foot-deep volcanic lake full of diluted sulfuric and hydrochloric acid. While local sulfur miners squatted on shore, looking on in silence, Maurice paddled happily around, now and then fighting a breeze that tried to push him to the middle of the lake. He took bottom samples until the acids ate through the steel cable hooked to his collecting bottle. Of course, the Kraffts arranged to catch most of this on film, and soon their fame, or maybe their notoriety, expanded.

"A volcano is like a wild animal," Maurice explained in the interview I saw. "He's blind, he doesn't see you, and you are sort of a crazy witch doctor who wants to understand his stomach problems. So you have to stay a long time and observe his habits."

Then Katia, smiling behind her outsized glasses, said, "You have all these noises all around you, like you're in the bowels of the earth. And compared to this giant volcano, you are just *nothing*. This is very nice to feel." I found myself nodding as I heard her say this, and I remembered the way that my early travels through the immensity of geologic time had helped to put my mortality in perspective.

And finally Maurice, with a twinkle in his eye, said how fantastic it would be to build a titanium boat and one day *put in* on a bright orange lava flow that had risen from the secret heart of the world. It would be the kind of flow they had so often seen gushing out of Pu'u O'o on Kilauea, one of their favorite volcanoes. They had bought retirement land just outside Hawaii Volcanoes National Park, so that even in decrepit old age they could edge their wheelchairs to the wall at the old Kilauea caldera and peek in. Maybe, as it had in Mark Twain's time, it would be boiling again.

Maurice loved to talk about that lava boat, and maybe not just because it made for good press. Off they'd ride down the long slope of a volcano, he and Katia. They'd fly the French flag and wave to all their friends as they navigated an incandescent river down to the sea!

In late May of 1991, the Kraffts talked by telephone with Harry Glicken, a volcanologist friend who was doing research in Japan, on the far western edge of the Pacific Ring of Fire. They talked about how Mount Unzen, on the island of Kyushu, not far from Nagasaki, was really starting to pop. A great number of small pyroclastic flows had been racing down the mountain's slopes in relatively contained routes.

Glicken had devoted his career to researching the behavior of pyroclastic flows. His story was well known in the world of volcanology. As a young man in 1980, during the weeks leading up to the final, dramatic eruption of Mount St. Helens, he had been living in a trailer on a forested ridge about six miles directly across from the trembling flank of the snow-covered volcano. With a pair of binoculars, he had watched the flank bulge outward at a rate of up to five feet per day. His job was to look for evidence of mudflows or avalanches and report his findings via radio to the USGS. St. Helens's last dangerous eruption had occurred more than a century earlier. No one knew when or how or to what degree its ongoing eruptions would increase. As a precaution, however, Forest Service workers had flushed hunters and fishermen from the woods near the volcano.

On the day before the eruption, Glicken left his post to drive to the University of California at Santa Barbara for a long-scheduled appointment with his geology professor to discuss his graduate work. As arranged, his place was taken by his USGS supervisor, Dave Johnston. The next morning at 8:32 A.M., the swollen flank of the volcano suddenly rippled, churned, and slid away in gigantic blocks. It was the start of the best documented landslide in history. Within seconds, the slide released the fantastic pressure that had been building inside the volcano. Out burst a pyroclastic flow with temperatures of 500 degrees Fahrenheit. The expanding cloud atomized or flattened almost every Douglas fir in its path, peeled the paint off a car and flung it across a road.

"Vancouver! Vancouver! This is it!" Johnston famously shouted into his walkie-talkie. Those were his last words.

Harry Glicken, of course, was spared, and ever since he's been studying pyroclastic flows. When he met up with Maurice and Katia at Mount Unzen, their passions for volcanoes and disaster prevention would merge.

According to a Japanese legend, the shake and rumble in volcanoes like Mount Unzen is caused by a giant catfish. Normally the gods keep the fish under control by pinning it beneath a large rock. But sometimes, when the gods are sleeping or away, the fish wriggles free. One day in 1792, the catfish beneath Unzen thrashed so hard that a landslide roared down over Shimabara City, at the foot of the volcano, and into Shimabara Bay. The tumbling mass of rocks, shattered trees, and soil, plus perhaps the shock waves of the quake, generated a tsunami, which took off across the bay in a great watery bulge. When the bulge rolled onto distant shores, it pulverized wooden homes and villages. The eruption killed 15,000 people in one great blow.

Then Unzen fell back asleep. A graceful, tiled-roof temple was built at the base of the volcano to commemorate the dead, and over the next two centuries a population once more grew out around the mountain. Tea and tobacco farmers planted brilliant green fields across the ash-rich slopes. Despite the temple and the presence of numerous hot springs, people began to forget or ignore the fact that they were living on an active volcano. Up until the last few decades, none of them could have known that some volcanoes can slumber 700,000 years or more between major eruptions. Volcanologist Robert Decker has written that a human contemplating a volcano is like a butterfly that lives only a few weeks contemplating a thousand-year-old redwood it has landed on. How could it know that the tree was ever a sapling, that it hasn't stood there like a pillar forever?

Now the fish under Unzen was thrashing again. The 5,000-foot tall summit puffed out fire and ash, grew a grainy millimeter or two. Its tea fields turned black and gray. Magma creeping up through the throat of the volcano had formed a lava dome, a bulge near Unzen's bare summit, which began to crumble away in up to thirty pyroclastic flows per day. That was an extraordinarily high number.

Maurice and Katia hopped onto a plane. Over the years, their volcano travels had made them more and more aware of the potentially fatal effects of volcanoes on large, nearby populations. That was an awareness that they could help spread, and so they'd been working on a safety video for civil defense officials. Unzen's pyroclastic flows would probably give them some of the last shots they needed. On Réunion Island, Maurice and Katia had once filmed such a flow at night. Their slow-motion sequence showed lit-from-within rock chunks tumbling and bouncing downhill like giant rubies and sapphires, an eerie avalanche of the planet's riches. The eruption looked far more beautiful than deadly.

The Kraffts' fame and the drama of the eruptions had preceded them to Unzen. On June 3, a couple of dozen TV and newspaper photographers and reporters who had gathered near the foot of the volcano followed Maurice, Katia, and Harry Glicken to a makeshift observation post partway up a slope. Two miles from the source of the eruptions, the Kraffts unpacked their equipment and started filming. The site allowed them to view the summit as well as look down into the curving Mizunashi River valley—a ravine, really—where the flows had been rumbling harmlessly by for days.

"I hope to see bigger ones than these," Maurice had said almost impatiently, but with a smile, on camera. Then he and Katia settled in to wait. A Tokyo TV correspondent wearing a yellow slicker stood with his back to the summit, positioning himself for his cameraman. Others fussed with shooting angles and light meters. They adjusted their lenses, tripods, and microphones.

The continents were adjusting, too. The Pacific plate was grinding itself beneath the Eurasian plate. Molten rock that had been edging up from deep inside the planet rose through harder and cooler rock, slipped up between the subducting plates, climbed toward Unzen's summit. More rock soup shoved from below, squeezing some of the magma to one side, making the lava dome bulge.

Finally the dome broke.

Squinting through his lens, Maurice probably saw right away how uncommonly large this eruption was. Katia, peering through binoculars, might have noted how fast the debris seemed to be coming. Down the valley along the path of the other eruptions roared the expanding ink-dark cloud shot through with rocks and searing gases and the swordplay of lightning.

"Civilization exists by geological consent, subject to change without notice," writes historian Will Durant. And without any notice at all, the searing avalanche—which turned out to be ten times larger than any of the flows that preceded it—split in two. One fork, dense and heavy, curved downslope as expected. But the other, lighter fork veered hard right and surged uphill. It took dead aim at the ridge where everyone stood.

At the nature and literature conference that was my nominal excuse for flying to Hawaii, I met a teacher from Japan who told me she had seen a live broadcast of this event. The correspondent in the yellow slicker, talking excitedly, stood in such close-up on the TV screen that it was difficult to see what was closing in on him. And because he was so intent on the camera, he had little idea what was about to happen, either. Suddenly there was a roar. He started to turn, his handsome profile swinging around, his mouth flying open. Then everything went black. The footage was never shown again.

As the pyroclastic flow swept toward him, Harry Glicken would have known that there was nothing to do. Deep time was about to collide with the present moment, just as it had for Dave Johnston.

The Kraffts would also have known that there was nothing to do. Perhaps, though, Maurice and Katia had time to catch each other's eye. Perhaps they managed to reach out and hold hands. Perhaps they saw a titanium boat coming for them from the secret heart of the world.

Alfred W. Crosby, "The Largess of the Sun" (2006)

b. 1931

In works such as *The Columbian Exchange* (1972) and *Ecological Imperialism* (1986), global historian and geographer Alfred W. Crosby—like fellow author Jared Diamond, whose 1997 book, *Guns, Germs, and Steel: The Fates of Human Societies*, won the Pulitzer Prize—provides biological and geographical explanations for why Europeans were able to succeed in what he calls the "Neo-Europes" of Australasia, North America, and southern South America. In his 2006 book, *Children of the Sun: A History of Humanity's Unappeasable Appetite for Energy*, Crosby argues that since first cultivating fire, humans have depended on tapping new sources of energy for survival—starting with muscle power and moving toward increasing reliance on fossil fuels. Crosby's history explores how an ingenious and adaptable humankind found ever more efficient ways to harness "concentrated sun energy." As Crosby observes, "We couldn't be more creatures of the sun if we went about with solar panels on our backs."

Discussion Generators

While this essay is in the alternative energy chapter of this anthology, Crosby makes clear the scale of the Sun's massive power and our undeniable dependence upon it. In other words, it is not merely an "alternative." Describe what he means by his title phrase "Children of the Sun." Why does Crosby rely on so many famous Western artists, athletes, and scientists in his examples? Are there additional examples that you might find even more compelling? How does he blend scientific accuracy with human history in this opening section to his book?

> *The Sun . . . contains 99.86 per cent of the mass of the solar system.*
> —*John Gribbin, cosmologist (1998)*[1]

> *I do good to all the world. I give them light and brightness that they may see and go about their business; I warm them when they are cold; and I grow their pastures and crops, and bring fruit to their trees, and multiply their flocks. I bring rain and cold weather in turn, and I take care to go round the world once a day to observe the wants that exist in the world and fill and supply them as the sustainer and benefactor of men.*
> —*Garcilaso de la Vega, El Inca (1612)*[2]

Great scientists, great artists, great athletes—Einstein, chalk in hand at the blackboard; Michelangelo, paint dripping down his arm, working on the ceiling of the Sistine Chapel; Lance Armstrong wheeling across the finish

1. John Gribbin, *Almost Everyone's Guide to Science: The Universe, Life, and Everything* (New Haven: Yale University Press, 1999), 169.

2. Garcilaso de la Vega, El Inca, *Royal Commentaries of the Incas and General History of Peru*, trans. Harold V. Livermore (Austin: University of Texas Press, 1966), 42–43.

line of the Tour de France—strike us as wellsprings of energy. But no human being, indeed no life form whatever, produces more energy than it takes in, or produces any at all by itself. All humans and all organisms are dependent on external sources for fuels. All are parasitic.

The two principal sources of energy upon which we earthlings depend are the upwellings of heat, magma, and gases from within the planet and the radiation from the Sun. The former empowers exotic organisms such as those that live by the hydrothermal vents of the deep ocean; they won't be mentioned again in this book because they have little direct influence on human history. The latter source, sunlight, is by every measure the greatest source of energy, the fuel of life, on the surface of our planet. Here we are all, in the words of Vladimir Vernadsky, the Russian geochemist, "children of the sun."[3]

The Sun is the center of a dust twirl of planets and lesser specks, and is also the center of human life and of the life forms on which humans are dependent. And no wonder: that star makes up very nearly 100 percent of the mass of the solar system. Our Earth is no more than a mote of debris left over from its formation. A million of our planet would fit inside the Sun.

At its core, where the environment is exponentially many times more hellish than Dante could have conceived, the pressure and temperature are so extreme that no solids, liquids, or gases can exist, only plasma, the uninhibited swarming of subatomic particles which is the fourth possible state of matter. There, nuclei (the relatively massive central points of atoms) of hydrogen undergo a transmutation that is the foundation of our lives. Despite being of like and mutually repellent electric charge (like similar poles of magnets repelling each other, north versus north, south versus south), they collide head-on because there is no way of not doing so. When four hydrogen nuclei collide and fuse, the result is one helium nucleus.

The mass of the one helium nucleus is 0.7 percent less than that of the four hydrogen nuclei, and the missing mass converts into pure energy. That is, per collision, a very tiny amount, but there are trillions of trillions of these collisions per second. Furthermore, the quantity of mass involved is

---

3. Quoted in Vaclav Smil, *The Earth's Biosphere: Evolution, Dynamics, and Change* (Cambridge, MA: MIT Press, 2002), 9.

not as decisively important as what happens to it. To measure that, we resort to Albert Einstein's famous formula for calculating the quantity of energy represented by a given mass, $E = mc^2$ (Energy equals the mass times the constant squared). Said mass may be small, but *c*, the constant, is the speed of light—*squared*. This is the awesome conversion which we replicate on a minuscule scale with our hydrogen bombs. It is no wonder that we are advised not to look directly at the Sun.

The energy created by hydrogen fusion in the Sun's core rises to its surface and blows out into space in all directions as light, the fuel of life. Approximately eight minutes later and 93 million miles away, the mote which is our planet receives a half billionth of this radiation. Half of that is reflected back into space or absorbed by atmosphere and clouds. The paltry remainder is the largess that made and makes life on our planet's surface possible, including the lives of Einstein, Michelangelo, and Lance Armstrong.

Life on Earth began an immensity of time ago and for millions upon millions of years thereafter was limited in its greatest extravagance to prokaryotes, one-celled organisms without nuclei. One of the most common of these was cyanobacteria, blue-green algae. They made their living by directly tapping sunlight. Our name for their chemical masterstroke is photosynthesis (from Greek words meaning "light" and "to put together"). Cyanobacteria contain a greenish substance, chlorophyll, which makes the alchemists' hyperbolized Philosopher's Stone seem feeble. Chlorophyll absorbs and harnesses the energy of light to split water and carbon dioxide molecules. This makes other molecules, simple carbohydrates, the fuel from which and by which yet other molecules essential to the functions of life are constructed.

At an inexpressibly important moment long, long ago certain enterprising prokaryotes or perhaps new-fangled eukaryotes (cells with nuclei) engulfed cyanobacteria and did not digest but incorporated and recruited them. These cyanobacteria lost all ability to exist independently and settled down forever to sinecures as distinct entities called chloroplasts inside their hosts. They are there still, almost all of their needs filled by their landlords, who in the aggregate are our plants. In return for lodging, the chloroplasts absorb sunlight (they will even migrate within cells to gain better access to the light) in order to produce and supply their hosts with simple sugar, the basic food of both plants and animals. Human food chains always ultimately lead down to plants with chloroplasts arranging themselves to catch the sun's rays.

In the process of photosynthesis the oxygen component of the dismembered water molecules is discarded and drifts away. This is the source of most of the oxygen in our atmosphere, which animals like us take up and, via slow combustion (the technical term is "respiration"), turn the food we have ingested into energy. We are thereby able to function, to build and rebuild ourselves, to stay alive from minute to minute. The ash of this sedate burning is carbon dioxide, which we exhale, repaying the biosphere for the carbon dioxide which photosynthesis has used up. The energy that respiration produces drives our muscle to do "work." In physics, that noun specifically refers to the transfer of energy from one physical system to another, as when I apply pressure to these keys to type this sentence.

We gain access to that force via "prime movers" or indirectly from entities that tap prime movers. The prime mover concept dates back to Aristotle, and to St. Thomas Aquinas, who posited a mover who is never moved, but moves everything else, i.e., God. Physicists have pared the concept down to mean any machine (using that word broadly) that converts natural energy into work. I utilize muscle, humanity's first prime mover (it taps the oxidation of carbohydrates, a natural force), to press down my computer keys.[4] The windmill, a later prime mover, taps the movement of air to do the work of turning millstones and grinding grain into flour. A standard nuclear reactor taps the heat produced by atomic fission, a natural process, to transform liquid water into steam to drive turbines to provide us with electricity. (We use electricity to drive locomotives, elevators, streetcars, movie projectors, computers, and so on. You could think of them as secondary movers if there were such a category.)

When we first started, we fueled our personal prime movers with the food we acquired as simple harvesters of wild plants and animals. After a long while we elevated ourselves to the status of complicated harvesters. We domesticated (negotiated alliances with) a few other species, as have, for instance, ants (as we shall see in chapter 3), so we could have sources of food,

4. Steven Vogel, *Prime Mover, A Natural History of Muscle* (New York: W. W. Norton & Company, 2001), is a good place to start on this subject.

hide, fiber, bone, and help within easy reach. We harnessed fire (as have no other creatures), tapping sun energy by igniting biomass created by photosynthesis. The burning of recently living biomass—wood, for instance—has continued on into our time. In the last two centuries we have also been burning immense, almost immeasurable, quantities of fossilized biomass from ages long before our species appeared. Today, as ever, we couldn't be more creatures of the sun if we went about with solar panels on our backs.

In the last half century our demand for energy has accelerated to the verge of exceeding what is produced and can be produced by conventional ways of harvesting sun source energy. We are refining those ways, which typically tap the energy holding molecules together, through combustion, for instance. We are also trying to domesticate the energy holding atomic nuclei together, such as is extravagantly released in fission and fusion bombs. Our sun-struck physicists are even leapfrogging back over photosynthesis and committing hydrogen fusion in their laboratories.

Pattiann Rogers, "The Power of the Sun" (1994)

b. 1940

Pattiann Rogers's twelve books of poetry include *Firekeeper: New and Selected Poems* (1994), *Generations* (2004), and *Wayfare* (2008). Her work has been recognized with the Tietjens Prize, the Hokin Prize, the Bock Prize from *Poetry* magazine, the Roethke Prize from *Poetry Northwest*, the Strousse Award from *Prairie Schooner*, five Pushcart Prizes, and an appearance in *The Best American Poetry of 1996*, among other significant awards. Rogers's poems often celebrate the natural world, displaying an intimate knowledge of astronomy and biology as well as a lively curiosity in questions of consciousness and spirit. In all of her books there is a quest for holiness and wholeness, an exploration of the human connection with the Earth—with the lives of animals and plants and more abstract natural systems.

Discussion Generators

In "The Power of the Sun," Rogers stresses the importance of recognizing that our existence—our very identity as living creatures—depends finally on our relationship with namesakes. What does the act of *naming* have to do with the possibility of human beings reimagining ourselves as users of solar energy rather than exploiters of other forms of energy that might be more damaging to the planet? How does the poet weave together reflections on the power of language with more material ideas about energy? And what do death, freedom, taste, and the erotic have to do with our appreciation of the sun's inspiring power?

> I think those who have its name
> are luckiest—most fortunate sun bear,
> sun grebe and sun bittern, sunflower,
> sun spider, sunfish, sunbeam snake.
>
> And the name is apt, for the sun bear
> rises, reveals his orange-crested chest,
> flares with huffings and hot, bellowing
> pronouncements over the quiet of damp
> liverworts, mosses and mangroves,
> just like the morning.
>
> The amethyst, green-headed sun
> bird sucks and shimmers his wings,
> perching on yellow petals and beams
> of buds, like summer sun. And the sunstone
> glows as reddish as a spangled dusk,
> and the sun spider is golden and swift,
> like a tight circle of sun focused
> through a glass, spiking its way
> across the grasses. Sunfish—rock bass,
> bluegill and pumpkinseed—are buoyed
> all over with flashing spines and shafts.
> They float and yawn light through
> and through underwater.
>
> Maybe the name comes first, the word
> a binding predestination fulfilled

only subsequently with a proper being,
as a sea wind wild and whipping
against a fractured cliff comes first
and then the cypress emerges, gnarls
and knobs itself to fit the shape
of that buffeting.

Think how the sun starfish can press
its limy rays outward, striving
to fill the possibilities of sweeping
combustion in its infinite name
all of its life.

And back, back inside that first final
locus of black beginning void, I believe
the sound of *sun* must suddenly
have been sung, and then, of course,
it had to happen.

2

Someone said my real name to me
last night. Someone whispered
to me, before I knew, the name he knew
me to become last night.

Sun breath he murmured into my mouth;
sun hands sun-caressing he said,
kissing the ends of my fingers;
he called me *sun voice low and catching*
through parting forest rain; ruddy
sun beaded nipples drawn and bitten
by his lips; *sun thighs* he whispered,
tasting with his tongue, *rocking slow*
with salt on the sea; light as solar wind,

self-luminous body he called me again
and again. And I rose easily to my name
as if over the rim of the earth at dawn,
exhumed and spacious, shining
in his arms, as boundless and blinding
and released in that given radiance
as death by its name could never hope
to be in all of its dark freedom.

Chapter Five

The Real Costs of Energy

Adrian C. Louis, "Nevada Red Blues" (1992)

b. 1946

Paiute poet, novelist, and journalist Adrian C. Louis was born and raised in northern Nevada, the eldest of twelve children. He graduated from Brown University, where he later earned an M.A. in creative writing. He has been the editor of four tribal newspapers, including the *Lakota Times* and *Indian Country Today*. He was twice nominated for Print Journalist of the Year by the National Indian Media Consortium, and he was a co-founder of the Native American Journalists Association. He has written ten books of poems and two works of fiction: *Wild Indians and Other Creatures* and *Skins*, the latter of which was produced as a feature film directed by Chris Eyre that premiered at the Sundance Film Festival in 2002. Louis has won various writing awards, including fellowships from the Bush Foundation, the National Endowment for the Arts, and the Lila Wallace–Reader's Digest Foundation. His 2006 poetry collection, *Logorrhea*, was a finalist for the *Los Angeles Times* Book Prize.

Discussion Generators

The New York Times Book Review has described Louis's poetry as "wild, sometimes foolish, sometimes poignant, often cartoony and sporadically brilliant. There's nothing politically correct about Adrian C. Louis." Indeed, Louis's poem "Nevada Red Blues," which first appeared in his collection *Among the Dog Eaters*, eschews political correctness in its harsh condemnation of white Americans' treatment of the state of Nevada, which has made its living on—among other things—gambling, prostitution, and nuclear testing. Could Louis's message be as powerful if it were couched in less confrontational language? What do you make of the Neruda quotation that serves as the epigraph for this poem—in what ways does the subject matter of this poem evoke "live fire" that "inhabit[s] you," and how does that effect reinforce the speaker's anxiety about nuclear testing? (Note that *Numa* is the Paiute word for "the people," and *Taibo* means "white man.")

"Where live fire began to inhabit you." —Pablo Neruda

We live under
slot machine
stars
that jackpot
into the black
velvet
backdrop
and
mirror the greed
of the creatures who soiled our land.

Numa,
it was
not
enough
for
Taibo
to make
our sacred land
a living
though
pustulous
whore.

He
had
to drop
hydrogen bombs
where
thousands
of years
of our blood
spirits lie.

Janisse Ray, "First Landfall" (2010)

b. 1962

Environmental activist and poet Janisse Ray is the award-winning author of *Ecology of a Cracker Childhood*, a work that combines elements of natural history and autobiography. Ray alternates chapters between her childhood in rural South Georgia and the ecological history of that region, demonstrating the delicate symbiotic nature of the landscape and the people who are connected to the land. Raised in her father's junkyard, Ray was surrounded by an impressionistic landscape of discarded vehicles, wiregrass, and mobile homes adjacent to U.S. Highway 1. In order to survive the clutter, Ray found solace in a growing passion for the longleaf pine ecosystem that had all but vanished long before her time.

Discussion Generators

Ray's profound identification with her home landscape informs her advocacy for such places as the island in the Gulf of Mexico described in the passage below. Ray's essay has become all the more timely since the 2010 Deepwater Horizon oil spill that covered 68,000 square miles in the Gulf. How can stories like Ray's help us to better understand the effects of offshore drilling on individual humans' lives? Is there a landscape that you feel similarly passionate about (and that might, conceivably, be imperiled by society's economic/environmental impacts)?

I am almost forty-eight years old, in what I hope is the middle of my life, walking a remote path on the east side of a wild barrier island, Gulf of Mexico, Florida. The day is in early January, and I am celebrating a new year where bald eagles nest, where red wolves have been reintroduced, where migratory birds land and feed after their long journeys, and where signs of Native American habitation, mostly oystershell middens, still abound.

Friends and I boated here this morning, across Indian Pass to a protected inlet with 360 degrees of wildness except for one lobster boat easing through Saint Vincent Sound, toward Apalachicola Bay. I walked alone around a spit and down a beach that faces the morning sun, keeping to a strand firmed by outgoing tide.

I can't remember being on a beach more wild. I pause to examine sand dollars and sponges. Sometimes uprooted trees, eroding majestically in wind and tide, block the way. I walk backwards against the calendar, against the years, becoming the young woman I have been, who had time for nature, who did not sit for long hours at a desk—and on backwards toward the naturalist and the explorer and then the native I do not remember being, although the ancient landscape stirs in me reminders.

My friends and I are the only people, I am sure, on this wild January island, and now I am almost a mile from my friends, down a wild beach. But the wild island can only be as wild as the Gulf that surrounds it, and the Gulf can't get away from people. Around me the beach is littered with human detritus—plastic buckets, Styrofoam crab floats, rubber gloves from the oystermen, plastic drink bottles, aluminum cans. Part of a dock has washed ashore, treated lumber with nails exposed. A lobster boat passes and I hear its deckhands calling to each other.

The things we manufacture and use away from this wild refuge wind up here anyway, in the one place they should not be. I stand and look out at the Gulf, my head full of terrible thoughts, thinking about an oil spill and what would happen to this place, warm Gulf waters unswimmable, beach contaminated, sea life extinguished. The thought of it is too horrendous to dwell upon. The worst of human civilization insinuates itself in the best of what we have left.

I see a woods path and go inland. I am far, far away from everything I don't like about the world, and I am in the embrace of that which I love most. The maritime forest is alive with magnolia and live oak, with tall cabbage palms. This is not a thin forest, not a sparse forest, not a new forest, not a forest to be taken lightly. This is an old forest on an old barrier island where red wolves have been calling.

As I walk I am remembering the young woman I used to be, who came alive in wind and sun.

These feelings are especially strong because the Gulf of Mexico was the territory in which I came of age. Here I first saw plovers nesting on beach sand. Here I saw a freshwater spring bubbling from the salty depths of the Gulf. Here I experienced wildfire. Here I made my first bird list and retrieved my first scallop, caught my first shark. Here I tasted smoked mullet and found an ancient pottery shard and sailed and identified an oystercatcher. Here, as I said, I came alive.

As I walk I am making resolutions. As I walk I am coming back from the dead. In the New Year, outside will be inside. I will spend more nights out of doors, the way I used to do. I will canoe more rivers. I will celebrate the high holidays of the sun, earth's calendar, with ceremonies that involve fire and no money. I will pay more attention to birds.

I will use less, stay home more. I will think about the consequences of my decisions on even the smallest menhaden in the sea. Along my path are the large bones of sambar deer. In an owl pellet (marked by white stains on leaves, below a level branch) I poke around and find the tiny skull of a rodent. I am happy in these old coastal woods, happy in the middle of my life, happy in this moment.

Christopher Morley, "Elegy Written in a Country Coal-Bin" (1921)

1890–1957

Humorist, playwright, poet, essayist, and editor Christopher Morley, together with his wife, Helen, moved to Roslyn Estates in Nassau County, New York—a place they called the "Green Escape," which would be their home for the next thirty-plus years—in 1920. At the rear of the Roslyn property, Morley built a cabin called "The Knothole," where he would produce most of his work for the rest of his life. The poetry of Walt Whitman provided powerful inspiration for Morley. Although Morley's use of form differs significantly from Whitman's—Whitman wrote almost entirely in free verse, whereas the following poem follows an ABAB rhyme scheme—one can see the influence of Whitman's social egalitarianism in "Elegy Written in a Country Coal-Bin," as well as Thomas Gray's pastoral vision in "Elegy Written in a Country Churchyard," which surely provided the inspiration for the title of Morley's poem.

Discussion Generators

The poem chronicles the social transformation that came with the replacement of traditional coal heat with cheap oil in rural America. Should relatively wealthy people or countries insist that others remain at a lower standard of living in order to slow petroleum consumption? Were today's issues of scale less critical one hundred years ago? Although published in the early twentieth century, Morley seems to anticipate what might be called "Peak Coal," the end of coal as a dominant source of energy—and yet, nearly a century later, coal is still widely used throughout the world. What can we learn from this poem about the difficulty of predicting (and implementing) transitions from one energy source to another? If you were to write a Peak Oil poem, anticipating a transition toward a new way to "bring [your] morning coffee to a boil," which of the energy sources described in this anthology would you select? Consider writing such a poem.

The furnace tolls the knell of falling steam,
  The coal supply is virtually done,
And at this price, indeed it does not seem
  As though we could afford another ton.

Now fades the glossy, cherished anthracite;
  The radiators lose their temperature:
How ill avail, on such a frosty night,
  The "short and simple flannels of the poor."

Though in the icebox, fresh and newly laid,
  The rude forefathers of the omelet sleep,

No eggs for breakfast till the bill is paid:
   We cannot cook again till coal is cheap.

Can Morris-chair or papier-mâché bust
   Revivify the failing pressure-gauge?
Chop up the grand piano if you must,
   And burn the East Aurora parrot-cage!

Full many a can of purest kerosene
   The dark unfathomed tanks of Standard Oil
Shall furnish me, and with their aid I mean
   To bring my morning coffee to a boil.

Dan Bloom, "To heal the world, to 'repair the world' . . ." (2014)

b. 1949

Dan Bloom, who coined the phrase "cli fi," is a freelance journalist in Taiwan, where he blogs at Cli Fi Central, a clearinghouse for books and films that deal with the topic of climate change. He has said that he has not taken a vacation from blogging in eight years. Bloom graduated with a degree in English from Tufts University in 1971 and has been working in Asia since 1991. In a recent blog entry, Bloom mentioned that he has not taken a vacation from blogging in eight years—indeed, that he has been online every day since at least 2008. Bloom's relentless, even obsessive, attention to new works of climate fiction has played an important role in bringing this burgeoning science fiction subgenre to the consciousness of mainstream media outlets, and Cli Fi Central, with its dozens of original posts each month, is testimony to the explosion of this important new body of work.

Discussion Generators

Bloom emphasizes in this 2014 post that he does not make money from writing on the Cli Fi Central weblog and that his passion for cli fi is a personal obsession and not a vocation. Indeed, much of the content written on blogs is produced by amateur enthusiasts rather than academic professionals. In what ways might Bloom's brand of authority be different from that of someone who has studied climate fiction in an academic setting? How might fictional texts such as movies and novels provide a perspective distinct from the more dispassionate voices of scientists and policy makers? Do you think Bloom is correct in proclaiming that cli fi is here to stay as a major Hollywood film genre? How does the concept of "tikkun olam," or a comparable idea from your own cultural tradition, figure into your own motivation to live lightly on the Earth and perhaps even to engage in activism?

It's not every day that *TIME* magazine recognizes what a lone word coiner has done by coming up with a new literary genre term dubbed "cli fi" that has caught on. Yet, the May 19 issue of the weekly magazine gave a quiet shout out and reporter Lily Rothman went even a step further in her summer movie preview headlined "Godzilla, 'Into the Storm' and More Summer 'Cli-Fi' Thrillers," gently pushing the emerging genre directly to the titans of Hollywood.

Maybe you are wondering how an independent dreamer with his head in the clouds most of the time came to be devoted to "cli fi," the various steps of trying to popularize it along the way, and why reaching *Time*'s national—and international audience—just might be important. Let me tell you how this all came to be.

To make a long story short, I did it my way. I have no office, no secretary, no funding and no sponsors. Over the last decade, I've taken a very strong interest—some might call it an obsession—in climate change issues and global warming. My wake up call was in 2007 when I read that years's IPCC climate report from the United Nations, and when I learned that the future of the human species on this dear Earth could very well be in dire jeopardy unless we stop our wanton ways and burning fossil fuels like there was no tomorrow, well, I suddenly realized maybe there won't be too many tomorrows for humankind.

But rather than sit around and complain, I decided to try my hand at using the Internet and the blogosphere to become a kind of PR climate activist and try to find ways to raise awareness of these issues.

First stop was an idea I called "polar cities" as safe refuges for climate refugees in the distant future, and in blogging about the polar cities concept, I found a novelist in Tulsa who was willing to try his hand at writing a novel about them. Jim Laughter—his real name, not a pen name—sat down and wrote "Polar City Red" for me, and all the credit and royalties go to him and not a cent to me. It's his book. And to help promote it, I sent out a series of press releases and oped columns calling his novel "a cli fi thriller." The word caught on, don't ask me how, but here we are three years later and *Time* magazine has recognized the term.

As you know, Hollywood has long shown an interest in climate-themed movies, from *Soylent Green* to Darren Aronofsky's recent *Noah* about a flood, yes, our flood. But as the world continues to warm up miniscule degree by miniscule degree and puts the very existence of the human species at a very grave risk. I began to feel that by setting up a literary and movie platform centered around the cli fi meme, novels and movies about climate issues, both entertaining and with serious messages just might help wake up the world. And our political leaders.

Time wrote: "Some say films like these [such as 'Godzilla'] are helping define a new subgenre: 'cli-fi,' or climate fiction. It's a timely subject for the summer [of 2014], given that the National Climate Assessment released May 6 found that the U.S. is already seeing the effects of climate change. Though the havoc in each film is wreaked in its own way, all of them use environmental destruction to raise the stakes."

I believe such current and future cli fi movies will succeed in helping audiences to confront environmental issues, much the way Nevil Shute's 1957 novel *On the Beach*—and the subsequent movie directed by Stanley Kramer—dramatized the horrors of nuclear war and nuclear winter and helped raise global awareness of the issues involved.

For me, it's a spiritual thing. "Without a vision, a people perish," I recall reading in my youth in western Massachusetts where I attended Temple Beth El's Hebrew School three afternoons a week after regular school let out and battled with the rabbis about all sorts of existential and religious issues that captivated me. I was always a bit of thinker, not a deep thinker, not a PhD thinker, not an academic thinker, but an often-obsessed thinker kid with ideas. That was how I grew up.

So in my own informal and unsponsored crystal ball, as I told *Time* in a phone interview, I dream of a new Nevil Shute who is going to arise somewhere in the world and with his or her cli fi novel is going to wake people up with a powerful story which will later be turned into an even more powerful movie.

So I believe that it's now time for Hollywood to go cli-fi, and I think some studio heads already know it's happening.

David Brin, a well-known sci fi novelist, has been following my work with the cli fi genre, offering me advice and suggestions along the way to *Time*'s story, and although he told me he considers cli fi to be a subgenre of sci fi, he also said he likes what cli fi might be able to do to help wake up the world in its own small way.

I've been doing this cli fi work for free, ever since 2008 when I first blogged about the term and began contacting media around the world to see if any reporters wanted to promote the term. Very few did at first. I got many rejections but I never took them personally. I soldiered on, undeterred, because I knew I was doing the right thing.

Not as a novelist or a movie director, since I am neither, but as PR maverick who works under the radar and never gives up. To have *Time* magazine recognize my work makes the long wait worthwhile.

I'm doing this work for free. I don't draw a salary, and I don't mind.

I'm not a trust fund kid, but I did have a father who left me an inheritance more important than money: *menschlekeit.* And in pushing the cli fi meme forward, as a way of paying it forward in gratitude for a wonderful 65 years on this planet, this is also my way of saying thanks to my dad, the late Bernie Bloom of Avenue J in Brooklyn, born in 1915 and gone in 2005.

My dad was a plumber, and in his own kind of way, a scientist, too, and he passed on his compassion for the world to his five children. I have his vision and his soul behind me, pushing me forward every day, egging me on, telling me to "never give up, whatever the odds."

Because who ever would have thought that my cockamamie idea of cli fi would catch on and end up in the *New York Times* and now *Time*? Not in a million years. But it happened.

And all I can say is "thank you Bernie Bloom, plumber extraordinaire, who fixed pipes—and more. You taught me that it was important to also 'repair the world'."

That is what my small contribution to the climate fight is all about: "tikkun olam."

Jerry Martien, "Now the Ice" (2006)

b. 1939

After working as a carpenter for twenty years, Jerry Martien became a teacher of creative writing and nature writing at Humboldt State University in northern California. His poetry has been published in the anthologies *Poems for the Wild Earth* and *The Geography of Home: California's Poetry of Place*. The book-length collection *Pieces in Place* (1999) brings together the subjects that have formed the basis of Martien's own life and work: the art of carpentry, the poetic imagination itself, and the beautiful Humboldt bioregion he has fought to protect for most of his adult life.

Discussion Generators

In the poem "Now the Ice," Martien meditates upon his own complicity, as a fossil fuel–dependent tourist, in the disappearance of the very glaciers that he and his wife have come to see. Martien speculates about what future generations will say about the choices our generation—and those who came before us—have made and continue to make. Do you agree with Martien's assessment that Americans live "by burning / their past / their future"? What good does it do to point this out in a poem? Explain the rhetoric of self-incrimination as it works in this text and consider which aspects of your own lifestyle you would describe in this way.

Thousands drive here
participate in their vanishing. Jenny and I fly
rent car
take each other's picture: pale greeny-blue lake
white water cascading

down cliffs
as background. Along the mountain flank
small figures
up and down the trail
pilgrims.

The Americans they will say
worshipped Nature
worshipped Money. A restless practical nation
so engaged in
business they lived by burning
their past
their future. At the park interpretive center
a young ranger assures us: not really glaciers
their effects
what visitors come to see.

Above the lake: remnant of a million winters
sculptor of mountains
using rock
clawed by ice
from rock. 150 glaciers when George Bird Grinnell convinced
T.R.: protect these peaks from mining
development. Crown of the continent
said Grinnell.

Visitors arrived in early 1900's by coal-burning locomotive
first gasoline-powered automobiles
to Prince of Wales Hotel
America's Alps
built by a business buddy of William Taft. He got: 160 acres of
Blackfeet land
$25 per acre. We get: reservation living
27 diminished masses of ice.

First week of August
Jenny and I up to our
freezing knees in it
binoculars
scanning the distant shore. White ribbons of water fall from the
crags
splash against rock
into thimbleberry
huckleberry. Soon grizzlies will come
down from the peaks. Mountain sheep on the high slopes
more tourists.

To the ranger I say
I'll tell my granddaughters
come see them soon
later
might be too late. See 'em while they're hot
he says.

no water so blue
as this blue of their melting

Marybeth Holleman, "What Happens When Polar Bears Leave" (2007)

b. 1958

Born in Cleveland, Ohio, and raised in the Appalachian Mountains near Asheville, North Carolina, Marybeth Holleman transplanted herself twenty-five years ago to Alaska, where she writes poetry and nonfiction essays and teaches writing workshops in the University of Alaska system. Her essays, poetry, and articles have appeared in *The North American Review*, *Orion*, *The Christian Science Monitor*, and *Sierra*, among other publications. Her radio commentaries have aired on National Public Radio, and her poetry has been nominated for a Pushcart Prize. She writes for nonprofit organizations on environmental issues, including predator control, global warming, oil spills, and polar bears. Her latest book, *Among Wolves*, was published in 2013.

Discussion Generators

The following essay, originally published in *ISLE: Interdisciplinary Studies in Literature and Environment*, is both personal reflection and thought experiment. If global climate change wipes out the polar bear population in Alaska, Holleman asks, what will this mean to people whose experience of Alaska is defined, in large part, by the presence of polar bears? What other distant and invisible changes might now be occurring on the planet as a result of our energy-related practices? Why should we care about the possibility that our lifestyle might be damaging ecosystems and cultures in regions of the world we will likely never visit?

As we stood in line at the grocery store with our bread, lettuce, and chocolate chip cookies, this is what my husband told me that made me turn away so quickly I was left dizzy:

His friend Chuck, a federal government biologist, was flying over the Arctic Ocean for the annual bowhead whale count. While scanning for whales, he saw polar bears, but in ways never before recorded. He saw females with cubs swimming in the sea with no ice pack in sight. He saw a drowned polar bear floating in a sharp blue sea, then another drowned bear, and another, no ice or land, just open water for more than 50 miles in every direction.

Not until it was our turn in line did I turn back around, wiping tears and focusing on the face of my son. He looked at me, and grabbed my hand in his.

It was my own fault Rick told me when he did. Standing there in Anchorage, 700 miles from the Arctic Ocean, I had pointed to the magazine rack, to the cover of *Alaska* magazine: a full-frame shot of a polar bear face. In the corner, over an erect ear dusted with snow, were the words, "Bound for extinction?"

Polar bears are excellent swimmers. Though males often weigh more than a thousand pounds, they can swim for miles, their broad front paws like paddles, their long necks and narrow heads reaching above ice-studded water. Polar bears are so aligned with water that they are classified as marine mammals; *Ursus maritimus* means "sea bear."

But they're built only to swim from one ice sheet to another. They are not made of blubber and fin like whales, or seals, or walrus; they are made of fur and foot; they need the polar ice pack.

They are ice bears.

Three years after that day in the grocery store, everyone knows that polar bears are drowning. So, when my husband tells me that, on his last flyover, Chuck saw dozens of polar bears stranded on a single drifting iceberg, I'm not surprised. When he says there was no ice pack or land in sight, and the bears would all most likely drown, I feel once more that abiding dismay, that upwelling of anxiety, and say to him: "I don't know what to do with that information."

"Yes," he replies, "lots of people feel that way."

In the last 40 years, Arctic sea ice has shrunk by nearly half. In the next 40 years, we may lose it all. The polar bears may lose it all.

As the ice pack thins and recedes, polar bears are forced onto land, paddle-shaped paws lumbering across tundra, where they must hunt less nutritious food, where they compete with brown bears who are more adept at terrestrial living.

Or they are left to swim and swim, in search of the ice that was once there, swim to exhaustion, to drowning.

I tell my friend Karin about the drowning polar bears; we talk about dying forests and rising seas and world leaders who do not take action. I am on fire; I want to know what to do. Her voice is even and soft as she both agrees, and disagrees.

A Tibetan Buddhist, a high school English teacher, and the mother of a teenage boy whom I have known since his birth, Karin is a compassionate woman; she teaches attentively, but doesn't expect the teenagers in her care to even have a chance to change the world. It will be too late, she tells me, for them to grow up and vote new leaders into office, and for these new leaders to stop the rising seas and mass extinctions.

But her own son, her one-and-only, lifelong friend of my son James, what is her hope for his life?

"That he become a monk," she says quietly. That he pray and transcend this world that is soon to desert him. This world, according to her religion, is merely an illusion created by our wanting minds.

"Of course," she tells me, "it's not that simple."

She plants lilacs in her yard and tends their new growth; she creates her own hospice care for a dying cat. But she is already loosening her hold on the world even as she gets up every morning, sends her son off to school, enters a classroom, and asks her doomed students to care about getting an A in English, to write a sonnet, to consider what an eighteenth-century novelist had to say about life.

Already some areas in the Arctic are warming ten times faster than the rest of the planet. Already Arctic sea ice forms later in fall, depriving coastal villages of the natural barrier from fierce storms, which now erode their shores and flood their houses. Of 213 Alaska Native villages, 184 face flooding and erosion.

Already the sea surges and rises, and the villages of Point Hope, Kivalina, St. Michael, and Shishmaref have begun plans to move their villages inland—refugees of global warming.

Already the ice breaks up weeks earlier in spring, constricting the time villagers have to hunt walrus and *oogruk*. Constricting hunting for polar bears, too; the earlier the breakup, the poorer the condition of the polar bears. Declining conditions are most pronounced in their southern range: the bears are becoming thinner; females are giving birth to fewer young; cubs are taking longer to wean; fewer cubs are surviving to adulthood. In Hudson's Bay, for every week that the ice breaks up earlier, polar bears come ashore 20 pounds lighter. Within a decade, females could become so small that they won't be able to bear young.

The polar bears shrink as their ice shrinks.

Driving to pick up my son one August afternoon, I had to turn on fog lights to see my way. But there was no fog, and the sky was cloudless. It was the third day of smoke so thick it hid the mountains around me, coarsened the throat, kept children inside and headlights on all day.

That summer of 2004 was the worst season of wildfires in Alaska's recorded history. Over 6.4 million acres of forest went up in flames. The temperatures were, for the sixth time that month, at a record-breaking high. The leaves of birch trees, which should have been a late-summer deep green, were brown and shriveled from a record drought, green gone a month early, fall's yellow skipped.

Record wildfires. Record drought. Hottest summer ever. Anchorage shrouded in smoke, the broad peninsula of roads and neighborhoods, downtown skyscrapers, unruly braids of lakes and parks and greenbelts, all cloaked. The snow-spiked peaks of the Alaska Range across the broad expanse of Cook Inlet, all the simmering volcanoes, gone from view. Even the familiar lines of the Chugach Range edging the city disappeared.

Enshrouded we sat in our cars, driving our SUVs here and there as if there was no connection between our driving behavior, our big-car boom, our life in cars, and the smoke, the fires, the extreme heat and drought. As if we could drive and drive, a lumbering beast unable to see disaster in time to change direction, drive and drive, and the skies would clear, the sun hanging just so above us, the world around us immune to the unchecked spread of our daily habit.

And I, who knew the truth of it, I, along with the rest, driving.

Writer and teacher Carol Bly says she most admires writers who can speak of beauty and horror together; who can describe a shimmering landscape, and then have some dark thing happen in that landscape; who can write of Eden, and the fall of Adam and Eve.

It's an old story, beauty and horror, and we're drawn to it, drawn to just a handful of tales, told again and again with different details, same endings. Stories of forgiveness, rediscovery, redemption; of joy, wonder, light emanating from—*surprise!*—dark, sorrow, despair.

Would I have remembered what my husband had told me if the Arctic ice held no allure? If the polar bears had just been beautiful, and not also doomed? If his friend had seen their shimmering whiteness against the glazed blue of multi-year ice, and not small cubs swimming with no ice in sight, not the body of a white bear, face down, in an azure sea?

Would it have mattered?

I was thirteen on the first official Earth Day. Same age as my boy now. After school, I walked the neighborhood alone, thinking of the planet and of my adult life before me. It was the first time I'd thought of the Earth as a living entity, as something I could affect. I scanned the sidewalks and roadsides, looking for litter. I picked up one soda can beside the road, all the litter I found that day. Just one, but I still feel the coolness of that thin empty container, see it glimmer in the afternoon sun, still savor the heart-skipping lightness I felt the rest of the day.

What reflected in that soda can—I wanted more of that feeling. I wanted to be of use.

But one soda can is nothing. Did no good. What does?

Ringed seals make up 90% of a polar bear's diet. Ninety percent.

In April 2002, the bodies of hundreds of drowned seal pups were found in the Gulf of St. Lawrence. The sea ice melted earlier, the ice floes disappeared sooner, and the mothers, who congregate on ice floes to give birth, had been forced to bear their pups in open water.

Global.

Warming.

I can't make sense of it. I can't read about it. I can't talk about it. It's too big. What—large, stronger, more frequent hurricanes in the south, more rampant fires in the north? Another shelf of the Antarctic fallen off, adrift, melting? Animal populations shifting northward, northern species like polar bears and lynx running out of room? What can I do about all that? Here—I'll recycle newspapers. I'll pick up litter in the park. I'll bring my own coffee cup to the latte stand. I'll even think about driving a smaller car, or a hybrid. Promise. OK? Good enough?

Face down. An azure sea. No. Not good enough. None of this. None of the two thousand things I can think of to do is enough.

I resolve to get up earlier every morning and meditate.

I resolve to practice loving kindness, to be generous and compassionate toward all things.

I resolve to want what I have, and no more.

To be grateful, and aware.

As I sit in the doctor's office, waiting to begin the treatment to rid my body of abnormal cells, to ward off cervical cancer, I take long, deep breaths and try to think good thoughts.

As I drive to pick up James from school, I count my blessings, even as rain falls mid-winter and melts the banks of white snow, the weather itself a grim reminder.

My doctor has given me a prescription for Flexerol to help with tension headaches. Take one or two at night, he says. One or two, you can take them indefinitely, no side effects, no addiction, no worries.

Look, I tell my husband, I can refill this three times. I like the way they help me relax, help me sleep. They're so nice for long airplane rides, especially if the plane is late or cancelled.

But the headaches do not go away. They do not even lessen. The pain spikes when I laugh, or roughhouse with my son, or run with my dogs.

I am also supposed to do neck traction every day, but I forget. I have a hard time doing anything for myself when the world around me burns.

My husband, a university professor who specializes in conservation, boards a plane at midnight and flies halfway around the world to help a distant impoverished chaotic country deal with environmental damage caused by oil. The irony of the jet fuel it takes to get him there does not escape him. We laugh as I say he's off to save the world, but beneath the laughter we're dead serious. He is driven to try, and it's what I love most about him. After he leaves, though, I wonder how much of our striving is simply a desire to *do*, to *have done*—that restlessness we feel when faced with despair, the way constant movement clouds truth, defies meaning. It's as if we, too, are swimming, swimming, toward a shore we cannot see but still believe is there, somewhere.

Why did I make my son get up and go to school an hour early for jazz band practice? Why do I keep making James try things, track or band or honor society?

He is a bright boy, *gifted* they call it now. But he does not want to go to school. He does not want to play trombone or basketball. He only wants to lose himself in computer games, pretend to be something/somebody/somewhere else.

Why does this confound me? Do I think he is immune?

If the world is burning around us, and there's nothing each of us can do, if all that's left is to pray for one's own vision to see past this illusion, then why in God's holy name would I push my son to do all these other things?

What is it like for him, for all the children, growing up now? When the American Dream of abundance and opportunity is so frayed and worn that, really, who believes it any more?

They say that generation Y, the label for my son's generation, are homebodies. These kids like to be connected to family and friends. They like computer chat rooms and cell phones. They work collaboratively and do not like to be alone. Is this a form of busyness, of denial? Or is it some precursor to a solution, water on the flames?

I wonder what we'll call the generation after generation Z. I wonder if we'll start over at A, or if we're not expecting to be around that long.

A picture on my desk shows my son at one year old, walking in the shallow end of a clear lake on a mountainside in Prince William Sound. He wears nothing but his diapers and a white T-shirt. His blond curls have not yet been cut, and they drape down over his shoulders, a cascade of light. The August day is sunny-bright, the lake's water is clear enough to see the stones at the bottom, the edges are laced in wildflowers and wind-sculpted spruce. But the ripples he makes as he steps through the water break apart his own reflection.

The polar bears images awaken me in the middle of the night. Sow and cub, swimming in circles, no ice no rest in sight. Scores of white bears crowded together, stranded on a melting chunk of ice.

An increase in spring rains is causing their snow dens to collapse, killing females and cubs; earlier spring breakup separates dens from feeding sites, farther than cubs can swim; starving adult males are resorting to eating the smaller females, cubs, and yearlings.

This cannibalism, says a local environmental activist in one of a growing cascade of newspaper stories on polar bears, is "global warming's bloody fingerprints."

Mothers and cubs first.

I've never seen a polar bear in the wild. I want to go to Churchill, or Barrow, I tell my husband. We have to go see them before they're gone, I say, frantic. I can't stand the idea of not ever seeing them for myself, even if it's at a garbage dump outside a town.

But worse, I cannot stand the idea of living in a world without them. I may not have seen them in the wild yet, but I've seen them. We all have—their images are everywhere. On every nature calendar's December or January. On Christmas cards. On commercials for soft drinks. On hockey teams' emblems. Everywhere, their faces, that whiteness punctuated by two dark eyes, the dark nose. We are enamored of them.

I remember learning, soon after I moved to Alaska two decades ago, that polar bear hairs are hollow, and it's the light filling the shaft that makes them look white—a reflection of snow. Those hollow hairs help regulate temperatures, acting as heat escape vents. Polar bears overheat at temperatures above 50°F.

I remember learning about how they crouch on ice by a seal's breathing hole and wait. How they hide their black nose with a paw, or clump of snow, making the disappearing act complete.

My son tells us one night, after dinner, that polar bear skin is black. This is something my husband and I didn't know. We look it up and James is right: their black skin captures heat while the hollow hairs release it.

Even if they do become extinct, which seems not just likely, but inevitable, we will still have the pictures. We'll still have that image of the white triangle of a cub sitting between the two front legs of the mother, sturdy pillars of protection. And the bear lying on his back, hind legs splayed in relaxation, paddle-shaped front paws limp and resting by his furred face.

Extinct, would they still appear in calendars? Would we still see their likeness in cartoons? Or would we banish their image as they have been banished? Would guilt and grief win out, or would aesthetic desire for beauty rule?

I've never seen images of other recently extincts—sea cow, passenger pigeon, dusky seaside sparrow—casually displayed in popular culture. But in this age of accelerated extinction, that might change.

Suppose an animal's extinction was rapidly followed by the erasure of their image from our grasp. Suppose we had to remember them only in our minds, our hearts. Suppose our struggle to maintain that fading image in our mind's eye was ultimately futile, as it is with the face of a lost beloved, always receding, always disappearing, no matter how much we want to hold on to it.

What I want to know is, what will it take for us to save the polar bears? What will it take for us to turn this around in time for polar bears, and ringed seals, and walrus? For generations Y, and Z, and A?

Tell me and I will do it.

The lord of lords, the one we all fall down on our knees to at one point or another in our small sweet lives, even she will not tell me. Supplication provides no relief, no answers. There is no solace for those who do not avert their eyes. And I will not avert my eyes. And I beg you to not avert your eyes.

If I were to ask you, if I were to ask everyone I meet, do you want polar bears to become extinct? To a woman, to a man, you'd say, no.

I have so many questions without answers. I fall down into the darkness of them. I have only images, handfuls I grasp and sift through, looking for answers by a light so dim it could be coming down a shaft of hollow fur.

There's a boy, sleeping. His mother tells him a story each night at bedtime. She has told him a different one, one she creates as she speaks, every night of his life. As he grows, she worries the stories are too simplistic and should become more realistic, like the books he is reading, where everything does not end happily and animals don't talk. But these new stories don't help him fall asleep. Instead, he tosses and turns for hours. He has nightmares, and stomachaches in the mornings before school. So she begins incanting the simple stories again. She returns to the world inhabited by animals who talk and learn and love the way she loves the boy, the way he loves her. And every night he lies still and listens; he says, "good story;" he falls asleep. Each night she describes a world that slips away each day. More and more her stories are lies, and she knows it, yet still one comes to her, every night, and lulls them both to sleep.

"Look twice: once for love, once for survival."

Patricia Hampl has faith, she says, not in what will or won't happen, but in lyricism, "[. . .] an authentic response to the world's impossible contradictions which seem to resolve themselves, finally, as beauty." It is the artist's work not to celebrate, she says, but to express wonder.

"And something terrible resides at the heart of wonder."

Golden birch leaves lie scattered over outstretched boughs of spruce, creating a forest of Christmas trees three months early. I point them out to my son.

"Yeah, pretty," James says.

It was difficult to pull him away from his new Sims computer game, and at first he trudged along, head down, angry with me. Now he watches our dogs stop to sniff a clump of Canadian dogwood as if the red leaves held some new story.

"I wonder what they smell," he says, "it's like they have this whole other way of knowing things."

I tell him of research about dogs who know when their owners are coming home—not just by time of day or hearing the car, but from miles away, as if they have another sense.

"Oh, yeah, of course they do," says James.

We climb the wooded hill, golden leaves drifting down around us, and my gait lightens.

"Even knowing that the horrible and beautiful are together in the world, we pass the threshold into something finer," says Linda Hogan. It's the very act of recognizing, through all the layers of one's being, that the horrible exists side by side with the beautiful, and that it's in this coexistence, this rubbing against each other, that life glows brightest. A fine and lovely line.

Wait. Don't drift off: we're really going to lose polar bears, all of them, in a lifetime.

Listen: since the drownings began, new seismic exploration and oil drilling, onshore and offshore of the Arctic Ocean, has reached an all-time high. Right now, as their ice shrinks, we are at fever-pitch to pull the very substance out of their ocean home that we will burn into the very substance that is destroying their ocean home.

I can't stop yet. I aim to write into an answer. Or at least a beginning, the dim outline of a shore. Can you see a shore? If I stop swimming, I die. So do you.

Fear-mongering. That's what Alaska's only congressman called reports of global warming. "I don't believe it is our fault," said Don Young. "That's an opinion. It's as sound as any scientist's." It's just natural causes, he said, like an ice age. And in the next breath he said, even if it is human-caused, "You're not going to turn it around unless everybody stops living."

In their second public hearing, a member of the State of Alaska's Climate Change Commission, after testimony on the effects of warming, asked, "What about human breath? It's a major cause of warming. Look it up on the web." He was serious. As was another member who said, "Well, we need to remember the many positive effects of global warming." It'll be a great boon to fisheries, she said, it'll be good for Alaska's economy.

Increasingly, when I talk with others about polar bears, they say, so quickly and with such serenity, "Oh, polar bears are toast, there's nothing we can do for them now."

How is it that we can move so seamlessly from denial to fatalism?

Not by flood, said the Lord, will I end the world next time. Not by flood. Maybe it will be by fire, then, heat rising, global warming making the continent of Australia so hot that parts of it, scientists predict, will soon be "like living in an oven."

What has protected polar bears up until now, states one report, has been their isolation, the relative lack of human presence in their sea ice world.

As it becomes more difficult to find and reach seals, polar bears are seen more often, and in larger numbers, near villages. They are especially drawn to whale carcasses left from fall subsistence hunts. In Kaktovik, on Alaska's Arctic coast, polar bears start appearing in early September, waiting for the whaling crews to begin.

Polar bears are shrinking, vanishing. They are desperate, and moving closer to us with each desperate step.

"Nothing in the world can take the place of persistence," claimed Calvin Coolidge. His world was such a far cry from ours today, though. Could persistence save polar bears, the ice pack, and generations yet to come?

Horror and beauty; wonder and terror; black skin and hollow fur.

It is a dark November evening. The snow came early this winter, then melted, now ice runs up and down the roads, up and down the trails.

James and I took the dogs for a walk yesterday, through spruce and birch to Goose Lake. The low sun shot light across the white expanse. A dusting of frost over old snow glimmered into the sky's deep blue. The dogs' gait loosened, from the nose-down straight-on intensity of woods walking to an open jaunt. The breadth of ice and light sent all four of us sprinting and sliding over the frozen lake.

James laughed. "Oh," he said. "I love this ice. I love winter."

Now in this dark I remember sunlight bouncing off ice, and I call my son to bed. It's a school night; morning will come too soon and too dark, and it will be a struggle, but we will open our eyes, we will arise, we will skate through the difficult, the uncertain, the lovely day.

Gary Paul Nabhan, from *Coming Home to Eat* (2002)

b. 1952

Gary Nabhan has an incredibly diverse record of accomplishment: nature writer, conservation biologist, ethnobotanist, wild food forager, and, in a new venture foreshadowed by the excerpt below, experimental desert farmer. With his wife, Lauri Monti, Nabhan has started a farm near Patagonia, Arizona, playfully called Almuniya de los Zopilotes (Private Experimental Farm of the Turkey Vultures). The interlocking roles of ritual, tradition, conservation, and celebration are all well represented in Nabhan's various projects. He currently serves as Kellogg Endowed Chair in Southwestern Borderlands Food and Water Security at the University of Arizona.

Discussion Generators

The brief excerpt below comes from a chapter of Nabhan's 2002 book, *Coming Home to Eat: The Pleasures and Politics of Local Foods*, and clearly points to the hidden energy costs associated with our dietary habits. What are some ways we might respond to this statement by recognizing and perhaps reducing the energy-intensiveness of our own food? In the larger chapter from which this passage has been taken, the author and his neighbors harvest local cactus fruit ("cholla buds") and prepare this simple food. What are the possibilities for local harvesting and preparation where you live? If trying to encourage readers to adopt an energy-saving lifestyle, what kind of language would you use in your writing?

The caloric cost of eating was not merely the number of calories produced by the fruit, nuts, meats, and roots we ate; it was also the effort expended in hunting and gathering, in processing and butchering, and most obviously, those we feel as heat while roasting, baking, grilling, or boiling our fare. Archeologists have estimated that most hunter-gatherers directly consumed a total of some 2,000 to 3,500 calories on the "average" day—including energy expended while carrying food to cookfires and gathering the wood used in grilling meats and tubers. Of course, averages didn't mean much to those who foraged in highly seasonal environments. And yet, when the Biodiversity Project newsletter recently published a discussion on my friend Peter Vitousek's estimate that most contemporary Americans require 46,000 calories each day to produce the food they eat, ecologist Stuart Pimm saw it and disagreed.

"Way low," he argued; Vitousek's estimate did not at all cover the many ways in which we consume fossil fuels to transport our groceries, supply our gas stoves, power our food processors and coffeemakers. What Peter and Stuart did agree on is this sobering fact: More than 40 percent of the earth's annual productivity is funneled into feeding just one species, our species, undoubtedly at the expense of the myriad other creatures trying to feed themselves on this wayward ark careening through space.

After the cholla bud harvesters left late in the evening, I walked outside, beyond my gate, and peered into the open pit where the cholla buds had been steamed, roasted, and smoked. The branchlets of desert broom in the bottom of the pit were charred, their aromatic oils reduced to a black tar. The heavy trunks of mesquite had been transformed into fluffy, grayish-white ash. The ashes and branches of desert broom rustled and stirred in the unrelenting wind. I heard in my head the echo of words said in my presence many times over the last forty years, but now I felt them etched into my muscles as well: "This is the body that has been given up for you. Take and eat it. Do this in memory of me."

John Updike, "Energy: A Villanelle" (1993)

1932–2009

Pulitzer Prize–winning fiction writer, poet, and critic John Updike is best known for his Rabbit series of novels. The series recounts several decades in the life of Harry "Rabbit" Angstrom, celebrating what Updike saw as ordinary America—usually meaning white, middle-class, small-town

America—yet also lamenting a decline into shallow materialism. *New Yorker* staff writer Adam Gopnik argues that "Updike's great subject was the American attempt to fill the gap left by faith with the materials produced by mass culture." Updike's attention to the rise of American consumerism helps to reveal how the United States lifestyle (including our heavy consumption of energy), which has been emulated by so many other societies throughout the world, has come to have such a deep impact on the biosphere.

Discussion Generators

Among the hundreds of Updike's essays, stories, and poems that appeared in *The New Yorker* is his poem "Energy: A Villanelle," published in 1993. In this villanelle, Updike explores our consumption of stored solar energy for comfort and convenience. How do the meaning and tone change with each repetition of his phrase "the cost grows higher"? Is there something about the villanelle form that is especially well suited to the subject of this poem? What are the various dimensions of "lifestyle cost" that we should keep in mind in relation to our use of energy?

The log gives back, in burning, solar fire
  green leaves imbibed and processed one by one;
nothing is lost but, still, the cost grows higher.

The ocean's tons of tide, to turn, require
  no more than time and moon; it's cosmic fun.
The log gives back, in burning, solar fire.

All microörganisms must expire
  and quite a few became petroleum;
nothing is lost but, still, the cost grows higher.

The oil rigs in Bahrain imply a buyer
  who counts no cost, when all is said and done.
The logs give back, in burning, solar fire

but Good Gulf gives it faster; every tire
  is by the fiery heavens lightly spun.
Nothing is lost but, still, the cost grows higher.

So guzzle gas, the sunless night draws nigher
  when Man's sheer soul shall keep him on the run.
The logs give back, in burning, solar fire;
nothing is lost but, still, the cost grows higher.

SECTION THREE

energy syllabi

Chapter Six
Nature and Human Values

James Bishop
Colorado School of Mines

Course description:
Nature and Human Values (NHV) focuses on diverse views and critical questions concerning traditional and contemporary issues linking the quality of human life and Nature, and their interdependence. The course examines various disciplinary and interdisciplinary approaches regarding two major questions:

How has Nature affected the quality of human life and the formulation of human values and ethics?
How have human actions, values, and ethics affected Nature?

These issues use cases and examples taken from across time and cultures. Themes will include but are not limited to population, natural resources, stewardship of the Earth, and the future of human society. This course is writing-intensive and provides instruction and practice in expository writing, using the disciplines and perspectives of the Humanities and Social Sciences.

Required texts:
Bishop, Lyndgaard, and Slovic. *Currents of the Universal Being: Explorations in the Literature of Energy*. Lubbock: Texas Tech University Press, 2015.
Farca, Holles, and Richman. *A Student Guide to Nature and Human Values*. Plymouth, MI: Hayden McNeil Publishing, 2012.
Other required texts posted on Blackboard

Student learning outcomes:

At the conclusion of NHV, students will be able to:

1) Understand major ethical theories and concepts and apply them to current and past debates on technology, resource use, and environmental issues
2) Read and think critically about course reading assignments and lecture topics; discover personal biases and values, diverse perspectives, and rhetorical strategies
3) Construct written and oral arguments about course topics that are supported by relevant experts and evidence
4) Find and employ relevant research to writing assignments; consistently cite use of sources in-text and in bibliographies
5) Develop written work through a process of drafting and revision to produce clear summaries, comparisons, and analyses of texts
6) Appreciate the context of the engineering profession and the impact of work on social, environmental, and ethical systems

Brief List of Topics Covered:

1. Identifying, analyzing, and writing arguments
2. Critical reading skills
3. Major ethical theories in the West and ethical codes
4. Current debates in energy, technology, and engineering
5. Researching and writing a paper with multiple sources
6. Lecture series on professional, personal, and environmental ethics

Grading Procedures:

All students in NHV write three major papers, as outlined in the course guide, which are worth 150, 200, and 300 points. NHV has a grade scheme based on a 1000-point scale: 650 points for major papers, 100 points for the common exam, and 250 points for in-class work (class discussion, quizzes).

Grading Scale out of 1000 points:

A = 930+	B+ = 870–899	C+ = 770–799	D+ = 670–699
A– = 900–929	B = 830–869	C = 730–769	D = 630–669
	B– = 800–829	C– = 700–729	D– = 600–629

Revision Policy:
Revisions are allowed on papers #1 and #2. If you wish to revise your paper, you are required to do two things: (1) bring your paper to the Writing Center and get tutoring help with your revision, and (2) submit your revised paper no later than one week after your original submission was returned to you. Your final grade on the assignment will be the average of the two grades you received.

Common Exam:
All NHV students will take a common exam on the lecture content and the common readings from the NHV textbook. This will take place during finals week and will be worth 100 points (10%) of the grade. Students are encouraged to take good notes throughout the semester on both lectures and readings because they will be valuable in studying and preparing for the exam. Questions will focus on defining terms and arguments made in lectures and applying important concepts from the course. Most questions are matching or multiple choice, and none will ask students to choose a "correct" ethical solution to a problem.

Outline of Course Topics:
Week 1: Course Introduction
Week 2: Introduction to Ethics
Week 3: Science and Engineering Ethics
Week 4: Environmental Ethics
Week 5: Reflections upon the Grid
Week 6: Water in the West
Week 7: Engineering and Wilderness
Week 8: The Paradox of Development
Week 9: Machines, Mountains, and Muscles
Week 10: Digital Technology and Dialogue
Week 11: Nuclear Technology and Dual Use
Week 12: Rocky Flats: A Nuclear Case Study
Week 13: Mining for Minerals
Week 14: Fossil Fuels and the Future
Week 15: Wind, Water, Earth, and Sun
Week 16: The Real Costs of Energy

Chapter Seven

Generators: The Literature of Energy

Kyhl Lyndgaard
College of Saint Benedict and Saint John's University

Course Description

Energy is arguably the most important issue of the twenty-first century. If you believe energy can be described solely in work units of BTUs, MPGs, and kWhs, think again, as the human generation of expression has long mirrored our use of energy. We will examine the cultural and historical roots of human conceptions of energy, quickly reading our way to the present day while taking both local and global views of this inescapable and essential topic. Poetry and nonfiction will be regularly read, and a unit on world petrofiction is included.

Required Texts

Bishop, Lyndgaard, and Slovic. *Currents of the Universal Being: Explorations in the Literature of Energy.* Lubbock: Texas Tech University Press, 2015.

Crosby, Alfred. *Children of the Sun.* New York: Norton, 2006.

Ishimure, Michiko. *Lake of Heaven.* Trans. Bruce Allen. Lanham, MD: Lexington Books, 2008.

Gessner, David. *The Tarball Chronicles: A Journey beyond the Oiled Pelican and into the Heart of the Gulf Oil Spill.* Minneapolis: Milkweed P, 2012.

Habila, Helon. *Oil on Water.* New York: Norton, 2011.

Munif, Abdel Rahman. *Cities of Salt.* New York: Vintage, 1989.

Negarastani, Reza. *Cyclonopedia: Complicity with Anonymous Materials.* Melbourne, Australia: re.press, 2008.
Wolf, Christa. *Accident: A Day's News.* Chicago: U of Chicago P, 2001.

Course Requirements and Assessment

1. Regular attendance, participation in discussions and workshops: 15% of your course grade.
Classes and weekly work are highly important in this course. I believe that sustained, consistent effort is every bit as important to academic inquiry as doing well on a small number of exams or formal essays.

2. Responses: 15% of your course grade.
For many classes, I will be requiring a brief response to that day's readings, or some sort of prewriting based on longer essay assignments. These responses are 250 words minimum.

3. Co-led class discussion: 5% of your course grade.
You have the responsibility to co-lead a single class discussion on a reading.

4. Energy Connections: 5% of your course grade.
You have the responsibility to find and present a recent article/show/film/etc. that connects to one of our texts.

5. Final Exam: 10% of your course grade.
This exam will take the form of a moderated discussion on the semester's readings.

6. Three formal essays: 50% of your course grade.
Energy Regimes (15%)
Petrofiction (20%)
Diverse Energies (15%)

Page counts for the essays are 5, 8, and 6 pages, respectively (and approximately).

Schedule of Readings

Part I: Energy Regimes, Past and Present

Week One: Updike, "Energy: A Villanelle" (*Currents*) Patricia Yaeger's article "Literature in the Ages of Wood, Tallow, Coal, Whale Oil, Gasoline, Atomic Power, and Other Energy Sources" (PDF)
Week Two: *Children of the Sun* (up to 58); Rogers, Lee (*Currents*). *Children of the Sun* (59–116); Lorde, Hadley (*Currents*)
Week Three: *Children of the Sun*: 117–166. Chapter One: "Reflections upon the Grid" (*Currents*)

Part II: World Petrofictions

Week Four: Chapter Two: "A Going Thing" (*Currents*). Selections from Chapter Three: "Grappling with Fossil Fuels . . ." (*Currents*)
Week Five: *Oil on Water* (1–145)
Week Six: *Oil on Water* (146–239)
Week Seven: *Tarball Chronicles*
Week Eight: *Cities of Salt* Ch 1–23: 1–165. *Cities of Salt* Ch 24–45: 166–310
Week Nine: *Cities of Salt* Ch 46–65: 311–454. *Cities of Salt* Ch 66–77: 455–627
Week Ten: *Cyclonopedia*—Incognitum Hactenus: ix–xx; Bacterial Archeology: 9–72; *Cyclonopedia* Exhumations: 75–109; The Legion: 113–142
Week Eleven: *Cyclonopedia* Tellurian Insurgencies: 145–177; Uncharted Regions: 181–191; Polytics: 195–221

Part III: Diverse Energies

Week Twelve: Selections from Chapter Three: "Grappling with . . . the Atomic Age" (*Currents*). *Accident* (1–113)
Week Thirteen: Chapter Four: "Alternatives" (*Currents*). *Lake of Heaven*: Intro through p. 158
Week Fourteen: *Lake of Heaven*: 159–338
Week Fifteen: Chapter Five: "The Real Costs: Participating in the Vanishing of Energy" (*Currents*)

Chapter Eight
The Literature of Energy

Scott Slovic
University of Nevada, Reno

Poems are part of the energy pathways which sustain life.
—William Rueckert, "Literature and Ecology: An Experiment in Ecocriticism," *The Ecocriticism Reader* (1978/1996, 108)

Energy is Eternal Delight
—William Blake, "The Marriage of Heaven and Hell," *The Poetry and Prose of William Blake* (1793/1970, 34)

"Nourish your ch'i"
—Gary Snyder (e-mail, 10 November 2005)

Course Description:
From the essential energy that constitutes the most basic definition of life to the combustion of fossil fuels, the splitting of atoms, and the generation of heat and electricity through an assortment of "renewable" and "alternative" means, the concept of *energy* is clearly at the center of what we call "literature and environment"—indeed, if construed broadly enough, this phenomenon may be at the heart of everything we think of as life and literature, as civilization. Without energy, what else is there?

In the field of literature and environment (and the scholarly apprehension of environmental literature and culture, which we generally refer to as "ecocriticism"), we are accustomed to using such rubrics as "place," "animality/physicality," "self/other," and "social/environmental justice," among others, to guide our evaluation and categorization of literary texts and other cultural productions. In this semester's class, we will boldly—and perhaps recklessly—try to define and identify a new category of environmental texts, those devoted directly or obliquely to pondering the meaning of energy in human life and more generally.

Some would argue that society's quest for new and continuing sources of energy—and the polluting byproducts of the energy we use—is one of the most urgent environmental issues in the world today. Our energy-intensive lifestyles, particularly in places like North America, are daily topics of debate for pundits in the popular media and for scholars in the natural and economic sciences. So where does literature—where do the humanities—fit in? If literary art can help us to understand our physical environments and our relations to other species more deeply, can this form of imaginative expression also guide us to contemplate the implications of our energy needs and consumptive practices? Our contributions of vital energy to the world? These are open questions—and exploring them is the reason for this course.

Course Goals:

1. To (begin to) define, theorize, and identify "the literature of energy"
2. To become more conscious of our own daily habits of energy use
3. To become familiar with the major energy-related issues of our time and consider how literature and literary studies factor into discussions of these issues
4. To refine scholarly skills through individual and group projects

Required Texts:

Bergon, Frank. *The Temptations of St. Ed and Brother S.* Reno: U of Nevada P, 1993.

Brower, Kenneth. *The Starship and the Canoe.* 1978. New York: Harper Perennial, 1983.

Gaines, Susan. *Carbon Dreams.* Berkeley, CA: The Creative Arts Book Company, 2001.

Gelbspan, Ross. *Boiling Point: How Politicians, Big Oil and Coal, Journalists, and Activists Have Fueled the Climate Crisis—and What We Can Do to Avert Disaster*. 2004. New York: Basic Books, 2005.

McKibben, Bill. *The End of Nature*. 1989. New York: Anchor, 1997.

McPhee, John. *The Curve of Binding Energy*. 1973. New York: Farrar, Straus and Giroux, 1994.

Smil, Vaclav. *Energy at the Crossroads: Global Perspectives and Uncertainties*. 2004. Cambridge, MA: MIT Press, 2005.

Weisman, Alan. *Gaviotas: A Village to Reinvent the World*. White River Junction, VT: Chelsea Green Publishing, 1998.

Student Work:

Keep up with reading assignments, participate actively in seminar discussions and group projects, prepare a weekly list of discussion topics (for each assigned text), do a personal energy accounting project, conduct an interview with someone knowledgeable about a particular aspect of energy, write a book review (400 words) about a recent book relevant to energy, and prepare either a PowerPoint presentation on some aspect of energy literature or a conference-length paper (c. 10 pages) on such a topic. Because this is an experimental course that will treat a rather amorphous, yet-to-be-defined field, we will work together as a team more explicitly than in most graduate seminars; group projects will include a bibliography of energy literature and an anthology proposal (including, if possible, an introduction and sample headnotes).

Course Schedule:

Week 1—Course introduction: toward definitions of "literature," "energy," and "(the) literature of energy." Examine small samples of "energy poetry" by Gary Snyder and contributors to the anthologies *Season of Dead Water* (responses to the Exxon Valdez oil spill in 1989) and *Atomic Ghost: Poets Respond to the Nuclear Age*. How are literary and academic communities now engaging with the issue of energy?

Week 2—Considering cold, hard facts—what are the world's energy needs today, the available resources (and alternative resources), and the global balances and imbalances of energy use and energy resources? Work with Vaclav Smil's *Energy at the Crossroads: Global Perspectives and Uncertainties* as our primary source of information on these issues. Due: preliminary, informal "accounting" of your own energy use (home use, office use, daily transportation, travel out of town, etc.). Write this up either as a list of "data" or as a story/essay about your "personal relationship with energy."

Work together on "interview template" for conversations with "energy specialists." Also, further refine categories of energy literature for group projects: bibliographies and anthology.

Week 3—How to make the topic of energy accessible and salient for literary audiences? Talk about John McPhee's *The Curve of Binding Energy* this week.

Week 4—No class.

Week 5—Making physics human—extend earlier discussion of McPhee by comparing his work with Kenneth Brower's *The Starship and the Canoe*. Due: brief (3–4 pages) interview with someone from the sciences, from a local power company, a gas station attendant, an oil or power industry representative, etc. Record the interview and write up a short transcript (including brief introduction) to turn in.

Week 6—Now it's only us—McKibben's foray into global warming and the philosophical, indirect approach to issues of energy. Discuss *The End of Nature*.

Week 7—Literary and journalistic approaches to critique and prescription—compare Ross Gelbspan's approach to energy/climate issues in *Boiling Point* with McKibben's approach a decade and a half earlier.

Week 8—A story of imagination and idealism—the counter-narrative of Alan Weisman's *Gaviotas*.

Week 9—Fictionalizing energy in Frank Bergon's *The Temptations of St. Ed and Brother S*.

Week 10—Who should enter the fray? The social obligations of scientists, writers, and . . . literary critics, as explored in Susan Gaines's *Carbon Dreams*.

Week 11—Presentation of book reviews—what are the latest books relevant to energy lit?

Week 12—Bibliography teams meet individually with me—and then convene the entire class for the final hour of the seminar.

Week 13—No class: Thanksgiving Break.

Week 14—PowerPoint and/or paper presentations. Work on bibliography and anthology together if we have time.

Week 15—Course summary—possibly additional presentations. Submit book proposal.

Suggestions for Additional Reading

Adams, Henry. *The Education of Henry Adams*. Boston: Houghton Mifflin, 1918.

Bacigalupi, Paolo. *The Windup Girl*. San Francisco: Night Shade Books, 2009.

Ballard, J.G. *The Drowned World*. London: Gollancz, 1962.

Bergon, Frank. *The Temptations of St. Ed and Brother S*. Reno: University of Nevada Press, 1993.

Berry, Wendell. "The Use of Energy." *The Unsettling of America: Culture and Agriculture*. San Francisco: Sierra Club Books, 1977.

Blake, William. "The Voice of the Devil." *The Marriage of Heaven and Hell*. London: William Blake, 1790.

Boyle, T.C. *A Friend of the Earth*. New York: Viking, 2000.

Brower, Kenneth. *The Starship and the Canoe*. 1978. New York: Harper Perennial, 1983.

Brownstein, Michael. *World on Fire*. New York: Grove Press, Open City Books, 2002.

Bryson, Bill. *A Walk in the Woods: Rediscovering America on the Appalachian Trail*. 1998. New York: HarperCollins, 2006.

Byrne, John, and Noah Toly. "Energy as a Social Project: Recovering a Discourse." *Transforming Power*. Ed. John Byrne, Noah Toly, and Leigh Glover. New Brunswick, NJ, and London: Transaction Publishers, 2006.

Callenbach, Ernest. *Ecotopia*. 1975. Berkeley, CA: Heyday Books, 2004.

Conrad, James A. "City of Destruction." *Making Love to the Minor Poets of Chicago*. New York: St. Martin's, 2000.

Crichton, Michael. *State of Fear*. New York: HarperCollins, 2004.

Davis, Mike. *Ecology of Fear: Los Angeles and the Imagination of Disaster*. 1998. New York: Vintage, 1999.

Easterbrook, Gregg. *A Moment on the Earth: The Coming Age of Environmental Optimism*. New York: Penguin, 1996.

Ellison, Ralph. "Prologue." *Invisible Man*. New York: Random House, 1947.

Gaines, Susan. *Carbon Dreams*. Berkeley, CA: The Creative Arts Book Company, 2001.

Gelbspan, Ross. *Boiling Point: How Politicians, Big Oil and Coal, Journalists, and Activists Have Fueled the Climate Crisis—and What We Can Do to Avert Disaster*. 2004. New York: Basic Books, 2005.

Ginsberg, Allen. "Plutonian Ode." *The CoEvolution Quarterly / Journal for the Protection of All Beings* co-issue (Fall 1978).

Haines, John. *The Stars, the Snow, the Fire: Twenty-five Years in the Northern Wilderness*. Minneapolis: Graywolf Press, 1989.

Heinberg, Richard. *Powerdown: Options and Actions for a Post-Carbon World*. Gabriola, BC, Canada: New Society Publishers, 2004.

Heinrich, Bernd. "Racing Fuel." *Why We Run*. 2001. New York: Harper Perennial, 2002.

Hitt, Jack. "The Hidden Life of SUVs." *Mother Jones* (July/August 1999). http://www.motherjones.com/politics/1999/07/hidden-life-suvs.

Hogan, Linda. *Solar Storms*. 1995. New York: Scribner, 1997.

Illich, Ivan. *Energy and Equity*. New York: Harper & Row, 1974.

Kingsolver, Barbara. *Flight Behavior: A Novel*. 2012. New York: Harper Perennial, 2013.

Kinnell, Galway. "The Fundamental Project of Technology." *The American Poetry Review* (July/August 1984).

Komanoff, Charles. "Wither Wind: A Journey Through the Heated Debate over Wind Power." *Orion* (September/October 2006). http://www.orionmagazine.org/index.php/articles/article/178/.

Koerth-Baker, Maggie. "What Does It Mean to Be Comfortable?" *New York Times* (January 25, 2013). http://www.nytimes.com/2013/01/27/magazine/what-does-it-mean-to-be-comfortable.html.

Kooser, Ted. "Late Lights in Minnesota." 1980. *Flying at Night: Poems 1965–1985*. Pittsburgh: University of Pittsburgh Press, 2005.

LaBastille, Anne. *Woodswoman*. 1976. New York: Penguin, 1991.

Le Guin, Ursula K. "Living on the Coast, Energy, and Dancing." *Always Coming Home*. 1985. Berkeley: University of California Press, 2001.

McKay, Don. "Ode to My Car." *Apparatus*. Toronto: McClelland & Stewart, 1997.

McPhee, John. *The Curve of Binding Energy*. 1973. New York: Farrar, Straus and Giroux, 1994.

Munif, Abdul Rahman. *Cities of Salt*. 1984. New York: Random House, 1987.

Nearing, Helen and Scott. *Living the Good Life*. New York: Schocken, 1954.

Neruda, Pablo. "Ode to Bicycles." 1954. *Selected Odes of Pablo Neruda*. Trans. Margaret Sayers Peden. Berkeley: University of California Press, 2011.

Olson, Sigurd. "The Way of the Canoe." *The Singing Wilderness*. New York: Knopf, 1956.

Postman, Andrew. "The Energy Diet." *New York Times* (October 5, 2006). http://www.nytimes.com/2006/10/05/garden/05green.html?pagewanted=all.

Rich, Nathaniel. *Odds Against Tomorrow*. New York: Farrar, Straus and Giroux, 2013.

Rifkin, Jeremy. "The Dawn of the Hydrogen Economy." *The Globalist* (2003).

Robinson, Kim Stanley. *Pacific Edge*. 1988. New York: Orb Books, 1995.

Rosser, Simon. *Tipping Point*. Cardiff, UK: Schmall World Publishing, 2011.

Saenz, Benjamin Aliré. "Creation." *Calendar of Dust*. Seattle: Broken Moon Press, 1991.

Sinclair, Upton. *Oil!* New York: Albert & Charles Boni, 1927.

Smil, Vaclav. "Peak Oil: A Catastrophist Cult and Complex Realities." *World Watch* (January/February 2006): 22–24.

Snyder, Gary. *Turtle Island*. New York: New Directions, 1975.

Steinbeck, John. "The Turtle." *The Grapes of Wrath*. New York: Viking, 1939.

Swenson, May. "Southbound on the Freeway." *The New Yorker* (February 16, 1963).

Weisman, Alan. *Gaviotas: A Village to Reinvent the World*. White River Junction, VT: Chelsea Green Publishing, 1998.

Williams, Florence. "A Mighty Wind." *Outside Magazine* (February 2007). http://www.outsideonline.com/adventure-travel/A-Mighty-Wind.html.

Wolf, Christa. *Accident: A Day's News: A Novel*. 1987. Chicago: University of Chicago Press, 2001.

Young, Louise B. *Power over People*. New York: Oxford University Press, 1992.

Scholarly Readings

Barrett, Ross, and Daniel Worden, eds. *Oil Culture*. Minneapolis: University of Minnesota Press, 2014.

Canaday, John. *The Nuclear Muse: Literature, Physics, and the First Atomic Bombs.* Madison: University of Wisconsin Press, 2000.

Clarke, Bruce, and Linda Henderson, eds. *From Energy to Information: Representation in Science and Technology, Art, and Literature.* Stanford, CA: Stanford University Press, 2002.

Farca, Paula A., ed. *Energy in Literature: Essays on Energy and Its Social and Environmental Implications in Twentieth and Twenty-First Century Literary Texts.* Oxford, UK: TrueHeart Press, 2015.

Gold, Barri J. *ThermoPoetics: Energy in Victorian Literature and Science.* Cambridge, MA: The MIT Press, 2010.

Kahn, Douglas. *Earth Sound Signal: Energies and Earth Magnitude in the Arts.* Berkeley: University of California Press, 2013.

Lord, Barry. *Art & Energy: How Culture Changes.* Washington, DC: American Alliance of Museums Press, 2014.

MacDuffie, Allen. *Victorian Literature, Energy, and the Ecological Imagination.* New York: Cambridge University Press, 2014.

Nisbet, James. *Ecologies, Environments, and Energy Systems in Art of the 1960s and 1970s.* Cambridge, MA: The MIT Press, 2014.

Schneider-Mayerson, Matthew. *Peak Oil: Apocalyptic Environmentalism and Libertarian Political Culture.* Chicago: University of Chicago Press, 2015.

Index

About the Editors

Scott Slovic is professor of literature and environment and chair of the English department at the University of Idaho. He has published more than two hundred articles in the field of ecocriticism and written, edited, or co-edited twenty-one books.

James E. Bishop is teaching assistant professor of liberal arts and international studies at the Colorado School of Mines, where he teaches courses in environmental philosophy, public policy, and engineering ethics.

Kyhl Lyndgaard is director of First Year Seminar and Writing Center at the College of Saint Benedict and Saint John's University. His publications include articles in *Great Plains Quarterly*, *Green Praxis & Theory*, and *ISLE: Interdisciplinary Studies in Literature and Environment.*